FENCING PARADISE

'Richard Mabey is a man for all seasons, most regions and every kind of landscape'
Andrew Motion, *Financial Times*

'Britain's greatest living nature writer'
The Times

'Mabey structures his book with a journey through the biotopia of the Eden Project in Cornwall, from where he darts off in history and geography to offer sparkling mini-essays on specific herbs and plants and more general matters botanical and ecological'
Steven Poole, *Guardian*

'Like being led through a wondrous landscape by a native guide. This isn't a green diatribe, nor another of the "plants that changed the world" genre. It's simply a great book'
Jo Bourne, *Geographical Magazine*

FENCING PARADISE

THE USES AND ABUSES OF PLANTS

Richard Mabey

Illustrations by Sue Hill

eden project books

TRANSWORLD PUBLISHERS
61–63 Uxbridge Road, London W5 5SA
a division of The Random House Group Ltd

RANDOM HOUSE AUSTRALIA (PTY) LTD
20 Alfred Street, Milsons Point, Sydney,
New South Wales 2061, Australia

RANDOM HOUSE NEW ZEALAND LTD
18 Poland Road, Glenfield, Auckland 10, New Zealand

RANDOM HOUSE SOUTH AFRICA (PTY) LTD
Isle of Houghton, Corner Boundary Road and Carse O'Gowrie,
Houghton 2198, South Africa

First published 2005 by Eden Project Books
a division of Transworld Publishers

This paperback edition published 2006

A catalogue record for this book is available
from the British Library.
ISBN 9781903919323 (from Jan 07)
ISBN 1903919320

Typeset in Nofret by
Falcon Oast Graphic Art Ltd.

Printed in Great Britain by
Cox & Wyman Ltd, Reading, Berkshire.

1 3 5 7 9 10 8 6 4 2

Papers used by Eden Project Books are natural, recyclable products
made from wood grown in sustainable forests. The manufacturing
processes conform to the environmental regulations of the country of origin.

Contents

CONTENTS

THE SECOND EDEN

A small island somewhere in the west

FIFTY MILES EAST OF LAND'S END is the world's largest and most artfully contrived indoor jungle. Just fifty yards from the entrance your clothes are already sticking to your back and palm fronds slapping at your face. I'm here for the first time, and shuffle into line with the visitors, trying to keep my eyes on the path in front. Overhead, the sky begins to vanish behind a green haze. Behind me a woman is muttering anxiously, worried about mosquitoes. The pool materializes like a mirage. It's no bigger than a garden pond, but suspended in the forest it seems like a lagoon spirited in from *Coral Island*. The concrete banks are draped with ferns – and, as faint as footprints on sand, with two discarded fig-leaf bikinis. 'Gone for a swim,' a handwritten note reads. 'Back in a couple of millennia. Adam and Eve.' Out over the water, the fruits of the tropical plum bushes dangle like ripe apples.

The centrepiece of the pool is a small islet of rock, with an inscription carved in copperplate: 'Although species-poor for their size, islands are rich in forms found nowhere else.' There is absolutely nothing growing on it. But working my way round the edge of the pool I come across a surfboard, hauled out on to the concrete beach like a seal, and pointed hopefully towards the islet. It is, of course, no ordinary surfboard. It's made from balsa, and encased in a mixture of hemp fibres and plant-based resins. The fin has been fashioned from green ash and driftwood. 'The challenge,' a sign reads, 'was to make a surfboard from sustainable materials. We hope this surfboard will lead the way.'

Lead the way to where? Further, you sense, than the wave-dashed beaches of south-west England. The surf is the peninsula's wild fringe, a turbulence that continually reminds you of the world stretching out beyond its shores. But this board on its cement beach hints at much broader horizons. The mock-up of a south-sea-island pool looks for a moment like an emblem of the planet and its predicaments: a tiny oasis of life, clasping its hopeful talisman, and gazing towards the future. That the whole installation is set in an even more elaborately fabricated island in the west of the British Isles simply underlines the message. Welcome to the Eden Project.

The symbol of an island set in troubled waters has surfaced repeatedly ever since humans began to be concerned about their place on the earth. In North America, many tribal creation myths were based on the belief that the earth was the top of a giant turtle's shell projecting out of the waters. The idea has been revived recently, and ecologically inclined writers have taken to calling America 'Turtle Island'. Gary Snyder's collection of poems of the same name won the Pulitzer Prize in 1975. In his introduction he joins the idea with our new sensitivity towards the fragmentation of people and nature on the planet:

> A name: that we may see ourselves more accurately on this
> continent of watersheds and life-communities – plant zones,
> physiographic provinces, culture areas; following natural bound-
> aries ... The poems speak of place, and the energy pathways that
> sustain life. Each living thing is a swirl in the flow, a formal turbu-

lence, a 'song'. The land, the planet itself, is also a living being – at another pace. Anglos, Black people, Chicanos, and others beached up on these shores all share such views at the deepest levels of their old cultural traditions ... Hark again to those roots, to see our ancient solidarity, and then to the work of being together on Turtle Island.

In the early Middle East, at the time the first agriculturalists were trying to make sense of their new power over nature, the formless island rising from the seas was a favourite theme of creation myths. In the area now occupied by Israel and Syria, the Genesis myth that reaches a climax in the story of the original Eden begins with the universe surfacing from the midst of the waters. In Egypt, Nu, the father of the gods, *was* the primeval water. With nowhere to live, he drew himself back, and out of the depths of his watery self rose a mound in the shape of a pyramid. The sun-god Atum stood on it, and it became the beginning of the world, which Atum then populated with his own seed. Perhaps for the first time, the creation was seen not as spontaneously generated, or as a gift of nature, but as an artefact, manufactured by beings who were themselves made in the image of humans.

The image has been powerful in science, too. In the 1960s the theory of island biogeography began to be developed. It demonstrated how all isolated habitats and wild places on the earth behave, ecologically, as if they were real islands, with the same problems of vulnerability to change and separation from sources of new life. The more habitats are separated from each other, fenced metaphorically, the less resilient they become. So as major habitats such as forests and wetlands shrink and become cut off from each other, their species lose the opportunity to move and adapt. The natural remedies of regeneration and colonization are stopped in their tracks. Local species, unable to migrate, become extinct. The whole ecosystem implodes, and evolution's inherent drive towards diversity is put into reverse.

And at last we have begun to see the earth itself as an island. That first astonishing glimpse of the planet from space showed a lonely oasis, pulsing with life but perilously finite. Lewis Thomas wrote:

> Viewed from the distance of the moon, the astonishing thing
> about the earth, catching the breath, is that it is alive. The photo-
> graphs show the dry, pounded surface of the moon in the

foreground, dead as an old bone. Aloft, floating free beneath the moist, gleaming membrane of bright blue sky, is the rising earth, the only exuberant thing in this part of the cosmos ... It has the organised, self-contained look of a live creature, full of information, marvellously skilled in handling the sun.

You know that when you step through the gates of Eden, you are being asked to suspend reality. You know that this is a world of illusion, where you will see not so much the earth itself as its image, focused through a lens; where your imagination will be teased with memories and allusions, and prompted towards connections that might never have occurred to you. Yet it is, at the same time, very real. The plants are far from illusory. They smell of pitch, balsam, Spanish markets, autumn ferment – solid things, actual moments. Their thorns draw blood. The bananas that hang from the trees in the dome are the same as you buy in a supermarket. The water on which they utterly depend is also real, and if there is ever a local drought, they will die as realistically as if they were in the wild. If Eden is, in one sense, a model of the world, it is also a microcosm, which must live by the world's rules.

I've been invited to write about the Eden Project, to make a response to it. Which, of course, is what every visitor makes, and I doubt that mine will be any less personal or fragmented. Like everyone else, I'm carried away here on waves of free association. I spot plants I've met on travels abroad, quite renewed because of their setting. I sneak seeds into my pocket, dreaming of my own garden. I catch a scent that whirls me back half a lifetime, and then search hopelessly for its source. I get bumped and tetchy in the throngs of people, and wonder what lesson that gives us about the state of the planet. Then, sitting out with them in the sun, adrift on wafts of rosemary, I quite forget the serious questions Eden wants us to address, and wonder if *that* is a subliminal message, too. I watch small children earnestly lecturing their parents, and carrying their litter to the correct recycling bin – then stopped in mid-sentence by a tree that soars beyond the craning of their young necks. Nothing here is ever quite finished, answered, *over*.

Eden describes itself memorably as 'a living theatre of plants and people'. It isn't an entirely original idea. In 1640, King James I's apothecary, John Parkinson, published *Theatrum Botanicum. The Theater of Plantes. Or An Universall and Compleat Herball*, which aspired to be an account of all the plants then known in Britain, and the roles they played on the stages of medicine and commerce. But Eden takes the idea further than this, in its deliberate dramatization of the interaction between people and plants. Its approach is the opposite of didactic. It's provocative, questioning, curious. You can have the formal tour if you wish, and hear the facts set out straight. But you are more likely to be caught – like those children momentarily forgetting their gravitas – by some improbable artwork: by a honey bee the size of a food truck, by a 'Mayeux Tapestry' detailing the murky political history of chocolate in a frieze of outrageous cartoons. Eden explicitly believes that the way to bring us closer to an understanding of the role of plants on the planet is through our imaginations, through art and play, through the revival of old myths and the creation of new.

This is what interests me about the place. This journal, covering three seasonal spells at the Project, will be a pot-pourri of responses, recollections and free associations sparked off by the exhibits. But winding through them will be a continuing reflection on the art and mythology of plants. I mean this in the broadest of senses. The Project sets the scene for a very generous take on the idea of plants in myth by being a kind of outsize legend itself. It is part fantastical entertainment, part real habitat, part teaching aid, part utopian model. Even as a purely commercial operation, which has helped the economic revival of this part of Cornwall, it is as much a symbol as a real company.

Throughout the long history of human interaction with plants, myth, science and politics have intertwined. Sometimes mythology has been a useful brake on the excesses of progress. Sometimes it becomes an excuse for bigotry and cruelty. Separated from its cultural roots it can degenerate into propaganda. Separated from just about everything else it can become, more agreeably, pure entertainment, something which Eden is happy to promote.

Yet myth is a continuous source of worry. Fables and stories about the *mana* of plants – about how they came to be, how we should behave towards them, why some cure and some kill – go back as far as human language itself. And their influence over human societies has been profound. Decisions about the types of crop we grow, feelings about the status and role of agriculture, beliefs about the fertility of the soil, have all been shaped as much by ancient myths as by practicality and science. Today, for instance, a conviction that plants are in some way intrinsically health-giving has enabled a multi-billion-pound cosmetic and food supplement industry to develop. No wonder, perhaps, that in scientific quarters the renewal of interest in mythology is viewed with suspicion. We have progressed, it's argued, only by shedding such fanciful and subjective interpretations of the world.

We're creatures of reason, and any return to myth is a dangerous backward step, an act of treachery against civilization.

Curiously this kind of worry isn't felt about art, in which the life and role of plants have been richly explored. Perhaps the self-conscious 'labelling' of artworks gives them a safe distance from reality, which myths, with their more ambivalent status, lack. Yet maybe this is to misunderstand the way that myths work, particularly with modern listeners and interpreters. It is far from clear if they are taken literally even among pre-scientific societies. The anthropologist Claude Lévi-Strauss has suggested that myths are not simple explanations or 'justifications', but attempts to overcome 'contradictions'. As such they are not alternatives to scientific truths, but a kind of arbitration between them and the realities of belief and cultural traditions.

This is clear in the kind of popular culture that continues to flourish around plants. My own findings in the *Flora Britannica* project suggest that modern plant 'mythology', in which family custom and children's play are important, freely mixes science, social meaning and personal feeling in an attempt to reclaim nature from its appropriation by specialists. Its absorption of, rather than opposition to, scientific evidence is crucial. The ancient reverence for trees, for example, is justified by every new piece of evidence discovered about them: the immense age reached by some species, their status as the most complex of all plants, their crucial role in the regulation of the atmosphere. The belief in the potency of smell as a vehicle of memory has been confirmed not just for our species, but for most of the living world.

Eden's use of this powerful network of associations as a way into a better understanding of the planet has a history of storytelling and image-making behind it which goes back thirty thousand years, to the time of the Palaeolithic cave paintings. But imaginative stories may be more than simply educational tools. Making these symbolic representations of nature is central to the kind of creature we are. Language-use defines us as a species, and both limits and enlarges us; it's through the net of words and imaginative constructs that we experience the outside world, just as much as through unmediated sense impressions. Our challenge is to use this gift to reconnect ourselves with nature rather than separate ourselves from it.

Yet in working through these stories and images, we have first to accept that the plants themselves are neutral. A single species can begin as a wild plant, be harvested by hunter-gatherers, domesticated by farmers, cultivated by slaves, cropped by profligate machines, processed into junk food, become the source of new drugs, catch its own diseases as a consequence of cultivation, be allowed to go wild again, form an ephemeral ecosystem, become extinct. The political and social context of a plant's use is part of its ecology. Agriculture is never morally neutral. Nor is an attitude towards plants that regards them as only of importance when they are useful for the human species. Humankind would have no useful plants at all were it not for the protean inventiveness of the wild. Our new respect for our fellow beings – the conviction that they have a right to existence simply because they *are* – must begin to extend to plants.

Mythology has always acknowledged our complex moral relationships with the rest of creation, and never more so than in the fable that gave Eden its name: the story of the Lost Garden, and of humankind's Fall from natural grace, a story whose shadow has scarcely diminished in three thousand years.

THE
FIRST EDEN

The burning bush

DOWN AT EDEN IN THE SPRING, what fills my eyes is the furze. The stumpy bushes line the approach roads, spill down into the quarry. The whole crater is rimmed with a halo of chrome yellow, a bunched, pushy, exuberant garland. It's the wild edge to this great centre of culture. 'When furze is in blossom, kissing's in season', the saying goes, though I've always felt there was an implied 'where' in the 'when'. Furze is, in the best sense, a vulgar plant, a bush not to be beaten about. A brash, unpretentious, festive opportunist, a haunter of commons and wastes and of the places people visit to be festive themselves; a plant neither overburdened with oppressively mystical folklore nor associated with elaborately artificial landscapes. When the great Swedish naturalist Linnaeus visited Britain in the 1730s, he is supposed to have fallen on his knees and thanked God when he saw his first furze-covered common. He was probably on Putney Heath, just outside London, but it's a mark of the fondness that is felt for the bush that just about every open space in southern England has also claimed the privilege of playing host to him.

Furze – though we always called it gorse – has been a companion plant in my own life. It grew on the common around my childhood home in the Chilterns (where Linnaeus' reverent presence was believed in implicitly), and around the house I lived in when I first moved to East Anglia. Standing now in this furze-ringed chasm four hundred miles to the west of my home, I'm struck by how deeply our lives are embedded in myth-rich plantscapes. For me as a writer they're my bread and butter. I spend most of my time in preserved fragments of old wildernesses, models for future idylls, pleasure grounds, instruc-

12

tional parks, nature sanctuaries, lands of lost content, healing gardens, nostalgic gardens, gardens for the body and soul, yearning efforts to catch paradise in a patch of ground. It's hard to think of any piece of land out there that isn't also a mental *hortus conclusus*, hedged about with symbolic meanings. The living skin of the land is our seventh sense, a vast repository of memory of what, as a species, we have felt and done and dreamed. We can't see the plantscape 'straight', as a bee might. Even in the wildest places we experience it through associations and allusions.

I'm on a hillside in Crete, just before my spring visit to Eden. It's the Mediterranean high-time, before the long sleep of summer. The slopes are covered with pink cistus and purple lavender, with spiny brooms, spiny spurges, spiny thymes. The whole vegetation is prickly, hostile, resinous, armed with oil-skinned leaves and bristling branches against the parching sun and ravenous animals. The thorns and thistles of Genesis. At my feet the orchids and irises are bedded in a barbed fence from the realms of higher geometry: *Sarcopoterium spinosum*, the thorny burnet. A cousin of our soft, cucumber-scented salad burnet, it has evolved a network of protective twiggery exactly like a sheet of chicken wire. The angles of the twigs all form perfect hexagons, so that the plant looks like a vegetable honeycomb. Why this pattern, instead of an illegible tangle of random thorns? Do we read it wrongly? Fantasizing, I wonder if the honeycombed wire lures in homesick bees as well as repelling goats.

Imagination's indiscriminate, undisciplined freedom is, of course, its

curse as well as the source of its energy. The fancy begets the worry-
ing real-world fiction. The insightful myth turns into the corrosive
superstition. In medieval plant lore the mandrake (which in Crete grows
alongside the chicken-wire plant) was regarded as an exceptionally
powerful herb. Its forked roots occasionally resemble a crude human
figure, complete with genitals, and so, by sympathetic magic, they came
to be regarded as an aphrodisiac, a cure for sterility, even as the source
of a potion for driving out demons. So analogous did the plant seem
to a human being that special distancing techniques were advised for
picking it, lest the gatherer be implicated in – and avenged for – a kind
of murder. The Elizabethan herbalist John Gerard was scathing about
the mandrake cult, and pointed out that the root was no more human-
like than the forked roots of lowly carrots and parsnips. In 1597 he
wrote:

> There hath been many ridiculous tales brought up of this plant,
> whether of old wives, or some runnagate Surgeons or physicke-
> mongers I know not ... but sure some one or more that sought to
> make themselves famous and skillful above others, were the first
> brochers of that errour I speake of. They add further, That it is
> never or very seldome to be found growing naturally but under a
> gallowes, where the matter that hath fallen from the dead body
> hath given it the shape of a man; and the matter of a woman, the
> substance of a female plant, with many other such doltish dreams.
> They fable further and affirme, That he who would take up a plant
> thereof must tie a dog therunto to pull it up, which will give a
> great shreeke at the digging up; otherwise if a man should do it, he
> should surely die in short space after.

14

As a matter of medical fact the mandrake, like many other members of the nightshade family, is toxic and narcotic. In classical times it was used as an anaesthetic during surgery. Its peddling as a nostrum by 'physicke-mongers' must have been a fearful and threatening business, especially as the even more poisonous roots of white bryony were often carved into suggestive shapes (or grown in moulds and sown with grass seed to give them hair) and then passed off as authentic mandrakes. Much of the macabre superstition associated with picking the roots – the fatal shriek, the haunting of gallows' sites – was propagated by professional gatherers and quacks, anxious to keep others away from what was a lucrative livelihood.

What is one to make of this kind of myth, comic enough in retrospect but alarming and dangerous when it was current? Perhaps that positive myths need to keep some kind of touch with the grain of nature, and not degenerate into fantastical propaganda. Otherwise a story that begins as a way of making sense of experience can come to be a substitute for experience, an excuse for not confronting the realities of the world.

And one other reservation. If the making of myths and images is the way our human minds reflect and cope with the inventiveness of nature, can it also become a self-conscious process that separates us from it? The lilies of the field don't toil or spin, we were lectured, but they don't spend time in worried reflection on their condition either. Is that partly what the myth of the Fall is saying? That gaining self-awareness – the fruit of the tree of knowledge – was what finally put paid to our instinctive engagement with nature, and led to our exile from the Garden?

Yet the myths of plants can be beautiful and expressive, even when they are patently nonsense. For the best part of two thousand years a story kept emerging from the depths of central Asia about a creature that was half plant, half animal. A zoophyte. A chimera, like the Green Man, whose mouth and nostrils foamed with leaves. The tale first appears in the same body of Hebrew texts as the Genesis and Garden of Eden myths. The creatures were called 'Adne Hasadeh' (literally 'lords of the field'), and grew at the top of a 'navel' issuing from the ground. They were like lambs, covered with wool, and had limbs and cloven hoofs and jaws. In central Asia they were the 'Borametz', the 'lambs' of Tartary. And from the top of their flexible navels, their stems, they would graze the grass 'within their tether'. When this was exhausted they died.

Later writers coloured in more exotic details. The Borametz were the favourite food of wolves. They bled when they were cut, but the flesh tasted of crab, or crayfish. Odoricus, a Minorite friar from a village near Padua, reported that they burst from shells (or wombs) that resembled gourds. Baron von Herberstein (1549) added the vindicating information that the local Muslims used the wool for the small caps they wore on their shaven heads, and as protection

for their chests. Such a sensibly utilitarian story, he argued, couldn't be a fiction.

But by the seventeenth century, educated people at least were less sympathetic to such extravagant legends. There was a mood of confident rationalism in the air, and Enlightenment scientists poured scorn on the myth. None of these fantastical travellers had ever seen the vegetable lamb themselves. It was clearly just an ordinary plant, badly reported, hyped up by second-hand rumours and Tartarian whispers. The few supposed specimens that had reached the West were, on inspection, either obvious fakes, contrived from dried sheep embryos, or toy lambs made in China by whittling the woolly rhizomes of tree ferns.

But these small shreds of hard evidence didn't really explain the details of the myth, or its widespread currency, and it wasn't until the

1880s that a credible explanation emerged. Henry Lee, Fellow of the Linnean Society, argued in *The Vegetable Lamb of Tartary* that the Borametz was none other than the cotton plant, which had probably been introduced to central Asia from India at about the time the fables began. Writing his foreword from the redoubts of the Savage Club in west London, he concluded that we could now 'recognise its forms and features under the various disguises it was made to assume by the wonder-mongers of the Middle Ages', and that it was 'a plant of far higher importance to mankind than the poltry toy animals made by the Chinese from the root of a fern'. It was a rousing verdict from the heart of Victorian self-confidence.

What is odd is that this byzantine myth should have taken root in Europe at all. Cotton was a familiar plant. It had been imported by southern Europe in classical times, and by the whole continent from the fourteenth century. By the late sixteenth century cotton fabrics were being made in Britain by Flemish weavers. John Gerard, in his 1589 *Herball*, was nonchalant about it: 'To speake of the commodities of the wooll of this plant were superfluous, common experience and the dayly use and benefit we receive by it shew them.' He tried to raise it in his garden in Holborn. It 'did grow verie frankly, but perished before it came to perfection, by reason of the cold frosts that overtooke it in the time of flouring'.

Yet the fable of the vegetable lamb thrived. The fact that the mythical plant was identical to a staple raw material wasn't spotted. Or, perhaps more plausibly, was overlooked. The myth was romantic, poignant, moral, a parable about natural economy and the inseparability of plants and animals. It featured an animal with an umbilical connection to the earth. A vegetable that was nectar to wolves. A beast that hatched from a pod. A creature which died when it had exhausted its food supply. It would be hard to devise a more pointedly ecological fable.

And on its long journey it had merged with the fable of Eden itself. In the flyleaf of the early editions of John Parkinson's great seventeenth-century horticultural manual, *Paradisi in Sole Paradisus Terrestris* (a brilliant bilingual botanical pun: it means Parkinson's park-in-the-sun) there is an illustration of 'Adam and Eve admiring the plants in the Garden of Eden'. There, among a fabulous collection of palms, apples, martagon lilies, cyclamens, tulips and a ravishing long-haired Eve, is an unmistakable vegetable lamb on its stalk, still with a lot to eat.

The importance of myths isn't to do with their connection with the 'facts', but with the insights they provide into larger truths, often about our relationship with nature. Their construction is as characteristic of our species as nest-building is for birds, and they seem to be an essential ingredient of our collective unconscious.

Consider, for instance, a deity from third-century Peru (a great cotton-using culture since 2500 BCE) which was believed to be responsible for guarding farms. He has some similarities to the vegetable lamb. In a contemporary vase-painting his hair is made of snakes, entwined in braids. Plants of various kinds are growing out of his sides and mouth. He seems a wild and improbable hybrid, but essentially benign. He is, suggests the biologist Lewis Thomas, an imaginary version of a real animal, a small weevil from the mountains of New Guinea (Thomas calls it *symbiophilus*), which lives in contented partnership with dozens of plants in the niches and clefts of its shell, which are rooted all the way down to its flesh. 'It is thirty millimetres long, easily over-looked, but it has the makings of a myth.'

The question that we should ask of nature myths – and of natural shrines, fables, symbols, images, works of art – is not are they 'right', but do they work? Are they elegant expressions of a good idea? Can we work out what they mean? Do they help negotiate the gap between our restlessly imaginative brains and the limitations of the real world? Do they touch our memories and dreams? Make us laugh, make us think? And, most crucially, do they do all this without serving the covert interests of the powerful or harming the interests of the weak (a category which, of course, includes the plants themselves)? Maybe they need, as assessors, not so much the severe scientist as the sympathetic critic, the interpreter, the messenger who – adding perhaps just the slightest ornamentation – ferries them further on.

Down in this cleft in Cornwall, I'm encouraged to think that this is the role I should aspire to in this book. A kind of botanical gossip. A jester, who may occasionally wield the pig's bladder. A teller of tales which in the theatrical spirit of this place are as likely to be sparked off by a spell of freak weather or a visitor's stray remark as by the great set-piece exhibits. As I've said, the Project already has the quality of a fable itself, like one of those creation myths that have the universe rising as an island out of the abyss. The stories it narrates – of crops whose energy doesn't rely on the long processes of fossilization, of cures locked in seeds, of the calamities that follow greed and exploita-tion, of the extraordinariness of a creation that could dream up bacteria capable of living in undersea volcanoes and then move on to the chrysanthemum – are themselves fables as full of yearning as the myth of the vegetable lamb.

Genesis

ONE OF THE EARLIEST WRITTEN CREATION MYTHS (it appears on tablets from the third millenium BCE, and pre-dates written versions of the Old Testament's Genesis story by two thousand years) arose from the heart of what came to be called the 'Fertile Crescent' in Mesopotamia. Nammu was the earth-mother goddess and her son Enki was lord of the earth and its creative waters. Enki's father Anu, according to a later Sumerian hymn, gave him the power

> To clear the pure mouths of the Tigris and Euphrates, to make
> verdure plentiful,
> Make dense the clouds, grant water in abundance to all
> ploughlands,
> To make corn lift its head in furrows and to make pasture
> abundant in the desert,
> To make young saplings in plantations and orchards sprout.

All this work was carried out by the lower orders of gods, who were endlessly involved in dredging clay for the irrigation canals and piling up silt to fertilize the arable land. There was dissent in the ranks, at the fact that god-liness did not carry the perk of idleness. So Nammu has the idea of creating humankind to help

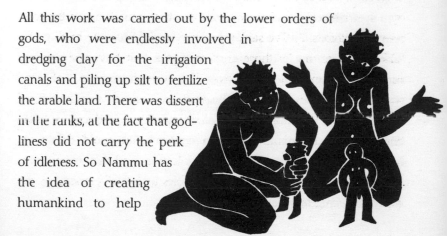

out with the work. Enki challenges her, and her sister earth-goddess Ninhursaga, to a competition to see which of them can fashion the finest men and women. They work them up from clay gathered from the edge of the primeval waters. Unfortunately, the gods, anticipating their release from toil, all get drunk, and the work goes badly. Nammu and Ninhursaga's creatures are weak, blind, maimed and incontinent. Yet Enki manages to allot roles for them, including the carrying of the soil baskets. But Enki's own creation, Umul, started in clay and finished in a woman's womb (and thus the first true human), is totally disabled. Ninhursaga cannot find a role for him, and in her anger she banishes Enki back to the region of Apsu, the watery abyss. Enki obeys, but demands that she looks after Umul: 'Sit him on your lap and praise him. Let men honour him until the end of time.'

There are features of this story which reappear throughout the mythology of Middle Eastern peoples, and which reflect the tensions of their life as agriculturalists. Water, the obsessional dream of desert dwellers, is ubiquitous. It is the source of creation, and of purification. It is yearned for so much that in some stories the tears of the gods flow into the irrigation canals. These gods, too, are ubiquitous. Where there had once been animal spirits, more or less equal members of the same cosmos as humans, there were now gods in human form, who had authority and dominion over them. Like the humans who invented them, they were unmistakably productive, managerial creatures, responsible for creating the world and organizing the life on it. It was their approval, not nature's, that now had to be sought.

One practical result of this shift was that the focus of fertility – originally located in the whole living system, then in specific agencies like rivers and the sun – moved, as agriculture began, inexorably towards the soil, the one element humans could influence. A later theme was that in the Golden Age of the first creation there was a state of perpetual spring. Farming peoples, forced to work out of doors through all the temperamental changes of the seasons, dreamed of a time when the weather was always clement, and the earth delivered its crops, without labour, throughout the year. Homer, writing in about 1000 BCE, described a perpetual spring in his garden of Alcinous, an orchard where the 'fruit never fails or runs short, winter and summer alike, it comes at all seasons of the year, and there is never a time when the soft West Wind's breath is not assisting, here the bud, and here the ripening fruit; so that pear after pear, apple after apple, cluster on cluster of grapes and fig upon fig are always coming to perfection'.

Agriculture is believed to have developed in the area now known as Iraq some ten thousand years ago. The idea of cultivating plants wasn't some sudden discovery, a blinding flash of inspiration by a hunter-gatherer. It was most likely an inevitable result of foraging for seeds and fruit, and the exploitation of the mass germination that must occasionally have occurred when these were stored or thrown away. There would even have been the beginnings of plant selection in this happenstance process. Plants with early or large or especially abundant leaves or seeds would have been preferred by foragers, and these characteristics would have been replicated in any plants germinating from food remains at their temporary settlements.

The only good evidence of what Stone Age Mesopotamians might have foraged for is the edible wild plants that are still used in the

Euphrates basin. Some have become extinct or have passed out of popular knowledge in the last few decades, but a large number of green plants are still gathered from the hills. The leaves of several mallow species are used in soups or stews with eggs. Salad plants include purslane, chicory, patience dock, watercress and the ubiquitous dandelion. More substantial are the edible roots: day-lilies, the wild onion *Allium victorialis*, and what is known locally as *taro* (*Colocasia esculenta*), a member of the arum family whose sweet tubers are eaten raw. There are also several leguminous plants with large seeds, some of which have been taken into cultivation, especially sennas (*Cassia* species), the heavy-fruiting pods of the carob tree, and wild peas and lupins. Almonds, pomegranates and chestnuts also occur wild, and maybe were the fruits in the minds of those who wrote the diet for the original Garden of Eden.

Most intriguing is what is known as *arouro* (honeydew). This is the sweet, sticky liquid excreted by aphids that have been feeding on the local oak trees. The oak leaves are picked and rinsed and the resulting solution boiled until it is a thick and aromatic syrup. This is a down-market version of manna, the secretions of insects that feed on tamarisk twigs, and which was the 'Bread of Heaven' for the Israelites during their wanderings in Sinai. Manna is still gathered by the Bedouin, who, like the Israelites, collect it early in the morning, when it is still cool enough to be solid, and pack it into pots. They make it into a kind of purée, and it is a precious source of sugar.

What modern desert foragers no longer gather are the wild grass seeds that were such a crucial agency in transforming the lives of their ancestors. The eventual growing of bulk quantities of species like *Triticum dicoccoides* – the wild ancestor of emmer and spelt wheat – was one of the foundations of the agricultural system that would come to dominate the world. The tendency of wild wheats to grow in large clumps, with their seed-heads all at roughly the same height, must have been strong cues in hastening the change from foraging to primitive cultivation.

But perhaps the development of irrigation was the decisive step. It was nothing much to start with, just the trick – familiar to any beaver – of diverting water for your own devices. But it changed things. The new irrigation channels became primitive field boundaries. They encouraged the separation and maybe proprietorial ownership of crops. Fixed fields made permanent settlements possible, and then inevitable. The growing of cereal crops which could be stored gave the new communities a measure of resilience during the winter. And as the combination of simple plant selection and irrigation began to push yields up, it meant that privileged and powerful members of society were freed from the need to forage for their own food. Almost every invention of civilization, from slavery and the division of labour to the development of the city and the birth of the accountant, followed.

There were huge accompanying shifts in people's perception of the natural world, too. Humans were no longer an unquestioning part of it, just one species going about its business like any other. They were now in a position to manipulate nature, and to have the illusion, at times, of dominating it. Nature came to be looked at not so much as the primary source of life and creativity, but as an object, a possession. No wonder the stories and fables of the Middle East reflect a bizarre and sometimes contradictory mixture of triumphalism, confusion and nostalgia.

How did the fables and myths that tried to explain these difficult transitions originate? It is not really until the beginnings of writing that myths begin to look as if they are being shaped by elders or priests, as political or religious propaganda. Before this, the pattern of their dissemination, and their frequent simultaneous occurrence in different cultures, suggests that they originally emerged from some shared, deep-rooted memory pool, and not just from the imaginations of shamans. In his study of peasant life, *Pig Earth*, John Berger has described the genesis and importance of another social phenomenon – gossip. This of course is a much more realistic and down-to-earth business than myth-making. But maybe these two kinds of storytelling have something of the same shape in the communal imagination. 'This combination of the sharpest observation,' Berger writes, 'of the daily recounting of the day's events and encounters, and of lifelong mutual familiarities is what constitutes village *gossip* . . . Indeed, the function of this *gossip* which, in fact, is close, oral daily history is to allow the whole village to define itself.'

The first written versions of the Genesis story (its oral roots are presumably much older) date from about 700 BCE, and appeared in the region of the eastern Mediterranean then known as Canaan. A local tribe of poor and dissident nomads, part herders and part short-term arable farmers, had rejected the extravagant fertility cults of the cities, and announced that they alone were 'the chosen people'. They had invented a single god in their own image – God the shepherd, God the planner of growth – and devised a myth to explain the creation by his hands. And perhaps with a wistful folk-memory of the luxuries of the hunter-gatherer's life, and a resentment at the inexorable grind of agriculture, they added a story to explain their condition. It was to become one of the most influential and beguiling stories on the planet: the fable of the Garden of Eden.

There are two versions of the myth in the written versions of Genesis. The first deals with the creation itself, and follows the form of many earlier Middle Eastern myths, with the 'firmament' arising from the midst of the waters. Except that in this case a single pre-existing god conjures the whole system into existence. Creation then proceeds in a proper evolutionary order. The earth is first populated by plants, then by fish, birds and mammals, the 'beast of the earth after his kind'. Finally God (rather tellingly slipping into the royal 'we') creates man (and woman) and establishes them in their role as agriculturalists, rulers of the rest of creation. 'Let us make man in our image, after our likeness: and let them have dominion over the fish of the sea, and over the fowl of the air, and over the cattle, and over all the earth, and over every creeping thing that creepeth upon the earth' – though, he adds, while adhering to a strictly vegetarian lifestyle: 'I have given you every herb bearing seed, which is upon the face of all the earth, and every tree, in the which is the fruit of a tree yielding seed; to you it shall be for meat.'

The second version, which introduces the Garden itself, is more subtle, more ambivalent. It takes off at the point when the creation of heaven and earth is finished, and then has God creating man in advance of other creatures. He forms him from 'the dust of the ground' and places him in 'a garden eastward in Eden' (perhaps a region north of Babylon, but really just a mythical oasis in a barren land). It contains 'every tree that is pleasant to the sight, and good for food; the tree of life ... and the tree of knowledge of good and evil'. Adam's responsibilities are to 'dress it and to keep it', eat what he likes, but stay away from the tree of knowledge, 'for in the day that thou eatest thereof thou shalt surely die'. Only then are the animals created, and brought before

Adam 'to see what he would call them', though the mate he was subsequently given, fashioned from one of his ribs, was at first called simply 'Woman'.

Then, of course, they eat the fruit of the tree of knowledge, and all hell breaks loose. Just what primal transgression this act of scrumping symbolizes has become a favourite puzzle of metaphysical thinkers. Some have laid the stress on Eve's eating of the apple, and attributed the Fall, literally, to greediness. Others, highlighting Adam and Eve's awareness of their own nakedness, suggest that the handing over of the apple was a metaphor for the beginning of sexual relationships. Perhaps it was humans trying to rise above their station – become gods themselves – or a matter of simple disobedience and the breakdown of authority.

But whatever the causes, the punishment dealt out by God was unambiguous and telling. Life was to become a hardship, a vale of tears. Women would bear children in pain and sorrow. And the unbidden harvests of the Garden would now have to be won by hard toil, *agricultural* toil: 'cursed is the ground for thy sake,' God rages, 'in sorrow shalt thou eat of it all the days of thy life; Thorns also and thistles shall it bring forth to thee; and thou shalt eat the herb of the field; In the sweat of thy face shalt thou eat bread, till thou return unto the ground . . .' Then, to stop matters getting any worse, he expels Adam (no mention of Eve) 'from the garden of Eden, to till the ground from whence he was taken'. It was a dramatic change from the gentle attentions suggested by that earlier duty of 'dressing and keeping'.

The new story of the Creation and the Fall doubtless had many agendas, but it is hard to ignore the agricultural motif. The Eden story justifies humanity's 'dominion' over nature, and then blames it for going too far. Farming is defined both as a salvation, a confirmation of human supremacy, and as a punishment, a penance for presuming too much.

Attempts to make amends for that aboriginal collapse, to recapture the lost ground and rebuild paradise, have shaped our relationships with the environments of the earth just as much as they've influenced the broad development of Western civilization. In the context of formal religion, the quest was for a while a literal hunt. Many of the early European explorers were spurred on by the conviction that they might discover Eden, the actual, physical Garden, somewhere in the far reaches of the globe. When they failed, others thought they might regain it by civilizing the new-found wildernesses; or that they might be able to reassemble, in some botanical Promised Land, the plants scattered across the earth at the Fall. The options through which humanity might regain grace on the planet proliferated: a return to a state of simplicity and harmony which echoed the original garden; the possibility that Eden might, on the contrary, be reclaimed by knowledge, labour, and physical as well as spiritual redemption. Even the wilderness itself – the spartan world of thorns and thistles in Genesis – had an appeal, as the environment humans had been exiled to, and from which they won their way back into God's favour. John Prest has written in *The Garden of Eden*:

> God lay closest to the simple life, even to the life of self-deprivation, and there was more virtue in a diet of locusts and wild honey than there was in all the luscious fruits of the Orient. This development of the belief in the wilderness as a place of trial and expurgation, or refuge and contemplation, the scene of covenanted bliss, led to the expression of a bewildering variety of contrasts and similarities between the Garden and the Wilderness. At times the challenge lay in creating a Garden in the Wilderness, and the terms employed are those of the pioneer. But at other times the Wilderness itself is considered to be the real Paradise, and the language employed is that of the Puritan.

Or as the American ecologist Andrew Little succinctly put it: 'Prophets do not come from the cities promising riches and store clothes. They have always come from the wilderness, stinking of goats, running with lice and talking of a different kind of treasure.'

Yet the allure of this vision of a lost paradise transcended orthodox religion. By the Middle Ages Eden had become a widespread secular symbol of innocence and hedonism, 'a delightful abode or resting place', 'a state of supreme happiness'. In *Richard III* Shakespeare dubbed England itself 'This other Eden, demi-paradise / This fortress built by Nature for herself'. In the 1960s hippies sang about going 'back to the Garden'. And in the 1990s an ex-rock music entrepreneur called Tim Smit decided to name his impossibly ambitious project to build the 'world's biggest conservatory' after the Lost Garden. 'What had begun as a project to exhibit plants from around the world and show we had domesticated them developed into a far deeper strategy to use plants as the common backcloth against which all human life is led'. Eden was an apt dubbing. Smit had already made his name through the rescuing of an abandoned Victorian garden in Cornwall. The Lost Gardens of Heligan became first a hit television series and a book, then a place of pilgrimage for tourists in the West Country. The new project would continue that initiative, and would include the use of a kind of ecological theatre to reinvigorate a local landscape and economy scarred by environmental thoughtlessness. They would build this Stadium of Plants in a worked-out china-clay quarry. The name Eden 'made sense', Tim Smit has written, 'as a symbol of Mankind in harmony with bounteous Nature. I also enjoyed the conceit that we had been thrown out of Paradise for eating from the Tree of Knowledge; perhaps only now, through the gathering of greater knowledge, could we return'.

It was also a timely name. The Middle East, where the whole story of plant domestication had begun, had long been another kind of theatre, for global feuding over oil resources and ideologies. In the year that Eden officially opened, the old region of Mesopotamia was on the brink of becoming the stage for a bitter war between Iraq and the West, in which the losers, as usual, were the ordinary inhabitants of all species, and their common environment.

Dramatic ironies

THE ROADS TO EDEN ON A spring holiday might be the approaches to Mecca at Ramadan. Except that this is a drive-in, secular shrine. You follow direction signs from ten miles out. Coaches This Way. Queues Possible. There are Eden teashops, an Eden bed and breakfast list, an Eden Gate internet café. It's already one of the most popular tourist attractions in Britain, with some 1.4 million visitors a year. Mostly they come because it is the most famous place in Cornwall, because to visit it, on the trail between the Lost Gardens of Heligan and Arthur's castle at Tintagel, has already become a tradition. Also because, like most tourists, they love gardens. On their way round this particular plot something, they hope, may rub off – the smell of rosemary, the name of a shrub to hide the shed, a few tips, maybe, for green fingers. And perhaps, momentarily, a feeling of duty done, a sense that, bathed for a whole day in such ecological wisdom, one may emerge a little greener oneself. The Eden Project is, like many communities of plants, a shrine. You come to pay homage and receive your blessing. In a village in Argyll, in the west of Scotland, there is a lone hawthorn by the side of a track. Local people use it as their wishing well, hammering coins into the bark, as if it were a slot machine printing out good luck mottoes in return.

The view of Eden from the portals, up on the rim of the old quarry, heightens its aura as a place of pilgrimage. The two great domes, fringed with prayer flags, one housing plants from the 'humid tropics', the other from the 'warm temperate' zones, lie in the very bottom of the pit like benign spaceships. Close Encounters of the Botanical Kind. They're constructed from hexagons, the old archetype. Honeycombs that nurture plants, thorny burnets cloaked in acrylic. They're called biomes. It's a word that sits comfortably on them, seeming like a dog-Greek contraction of bio-domes (though in fact it's a neologism for the biological communities in a particular climate zone, and refers, strictly, to the vegetation inside the domes, not the structures themselves). Every kind of signal – slope, gravity, perspective, amazement – drags your attention down towards them. They exert a centripetal force, and you need to drag your eyes outwards to take in the geography of the place.

Circling up and out from the biomes, climbing the walls of the quarry like cultivation terraces, are the supporting shrines, the sideshows. The notice in front of me lists them: 'Apples. Wheat. Lavender. Plants for Cornish Crops. Sunflowers. Tea. Eco-engineering. Hemp. Plants in Myth and Folklore. Imagine Building a World from Nothing...' Imagine re-building it. From here it's maybe six hundred metres to the other side of the quarry, and when I screw up my eyes the exhibits seem to assemble themselves into a pattern. This is partly deliberate. The beds have been sculpted in 'organic shapes' to follow the very unnatural contours of the quarry. And the congenial example of the English allotment, with its social as well as vegetable mosaics, has been an inspiration. Rows of plants are hatched in herringbone formations that cut across the slopes. A cascade of silver-foliaged shrubs tumbles down across the paths. In the distance more tracks, busy with builders, wind seamlessly out of the gorse thickets on the moors. There are fragments of stone walls, field gateways, the funnel of a train, a willow shelter, more flags, plants ranged on banks, then in echelons, down to the spick-and-span vegetable beds on the margins of the biomes. When I'm a little further down the main path I can make out gaudy ribbons of yellow courgettes and rainbow chard displayed between the rows. 'Explore your world with fresh eyes', the notices read. 'Could you survive without plants? How many have you used today?' Beyond conjuring up the ancient allure of the Lost Garden, Eden is also a temple to the idea of the plant as *solution*.

The Greeks had a word for this. The daughter of the god of medicine, Aesculapius, was called Panacea, and her name came to stand for any miraculous plant or substance that would be a universal remedy, a heal-all. Alchemists and early physicians spent their lives searching for it. Can we survive without, metaphorically at least, discovering it ourselves, without recognizing our dependence on the plant world? The parallel question is one we shy away from, because the answer is so humiliating: can plants survive without us?

Next time, hoping to beat that pull of the centre, I cycle in, towards the undecided land on the quarry's rim, where Eden merges with the real world. I'm based in a B&B a mile to the west, and I whirl through a labyrinth of narrow lanes lined with Cornish hedges. They are, I'm thinking, the most beautiful things in England at this moment in May. I freewheel past brocades of red campion and white garlic and bluebell, past sprays of lime-green fiddleheads, and notice from the corner of my eye the way the little round leaves and butterscotch spires of navelwort come and go as I move in and out of the shade. It's a real enough prospect, but a dream, too. This is the romantic vision of old England, a rural paradise where humans and nature lived in harmony, where plants were in their right place, at the right time.

Yet these aren't hedges in the sense that's understood in the rest of England. They're not topped by rows of blowsy hawthorn or studded with timber oaks. You'd have no luck trying to count the shrubs in a thirty-yard stretch to measure how old it is in centuries. (There aren't any.) The Cornish hedge is an artificial bank, a vertical rock garden, knitted together from slates and granite chunks and slabs of turf. Inside farmland the hedges are field boundaries, but along these lanes they exist to stop the fields from falling into the road. They're artefacts, yet somehow seem to have conjured together all the brimming wildness of woodland clearing and stream edge and cliff-top in one place.

Cycling past them is too broad a take. You catch the blocks of colour, the alternations of shade, little spikes of vivid, odd detail. But not the minute particulars of why such pieces of functional engineering should have a convincing patina of naturalness. The next day I walk in, and peer closer.

Bluebells, shield-ferns, lady-ferns, wood sorrel – plants I know from the deep shade of ancient woods in the south-east, plants which suggest they need deep leaf-litter too. But here they're out in the sun, with their roots wedged between stones. I draw my finger through one of the clefts, thinking of the niches on that little weevil, *symbiophilus*. I pull out a crumbling wad of dead moss, a curled-up millipede, flecks of stone and a smidgeon of mould. What on earth are the plants living on? Even making allowances for the balmy moisture of the West Country, which enables almost anything to grow anywhere, it's clear that, in the stonier banks, soil is pretty much a luxury. We have become fixated on what we reverently refer to as 'the fertility of the soil', as if the growth of plants was like the rising of cream from the ground. Maybe it's another relic of the first farmers and their fertility cults.

What was essential for the growing of mass-produced crops came to be thought of as necessary also for growth in the wild. The economies of nature couldn't be seen as cannier, better managed, more frugal than the economies of its self-appointed masters. The air-plant, clinging to the treetops and living on breezes, has always been an affront to the turnip.

But wild economies are more frugal. Soil in the horticultural sense – rich, deep, nurturing humus, the stuff that builds roots (and has become a metaphor for them) – is pretty much a creation of agriculture. Deep soils can build up from silt in river valleys and flood plains, but in most ecosystems, capital tied up underground is just so much wasted energy. Nutrients need to be out there, circulating in the organisms, mobilized in root systems and sunlit leaves.

Soils in the more strictly geological sense are, of course, ubiquitous and essential. They are the physical rootholds for plants, reservoirs for minerals and sugars, whole ecosystems in themselves. They contain complex communities of crucial organisms that recycle nutrients and keep the local chemistry beneficial to growth. But very few are the basis of 'the fat of the land' – 'fertile' in the sense that farmers understand the word. To continue with the economic metaphor, they are mostly short-term, rapid-cycled current accounts. They are often very thin but, crucially, adequate for the species that have evolved with them. When these lean groundbases are fertilized it is these indigenous species that suffer, crowded out by nitrogen-hungry crops and their associated weeds. A sandy heath or a chalk down (fifty species to the square foot) has only a covering of humus. An 8,000-year-old European woodland has maybe six inches. A tropical rainforest – the most complex ecosystem on earth – no more than a couple. And in the gaps between the stones in a Cornish wall there is just enough to coat your finger.

Beyond the lane, on the outer perimeters of the quarry, is Eden's back-lot. Security huts, think-tanks, unloading bays, workshops. I follow the service road down into the pit and then climb up to the inside edge, back into wild Cornwall. Or, more exactly, 'Wild Cornwall', a mocked-up exhibit of the county's green fringes, from which Eden has surfaced like an ambitious whale. This is a landscape of pure artifice, which has the feeling of one of those halls of receding and mutually reflecting mirrors. There are rows of carefully planted heather sprigs. There are the first shoots of a tall local rarity, Babington's leek, which grows in a few stony places near the sea. But other elusive West Country plants – Plymouth pear, bastard balm, tree-mallow – are there only as images, carved memorially in stone blocks at the foot of another stretch of wall, a reminder perhaps of their precarious existence. Early or English gentian is figured in the leaves of a stone Red Data Book. This diminutive lilac-flowered species is one of England's very few endemics, and it grows, in quite unpredictable quantities from one year to the next, chiefly in chalk or limestone grassland. In the sand-dunes of north Cornwall there is a unique April-flowering hybrid between it and the autumn gentian.

Next to it, and not more than a hundred yards as the crow flies from the Prideaux and Luxulyan lanes I've just walked through, there is a replica of a Cornish hedge. It's been made in the local vernacular, granite on granite, and planted up with foxgloves and ferns. Already opportunist species are edging their way into the cracks: dandelions, campions, wisps of grass. In fifty or so years it may well be indistinguishable from the hedges along the lanes. Except of course that it *is* different. It lacks pedigree. It was made self-consciously, for show, not utility. It's not exactly a fake, but it is a facsimile, a reproduction, and lacks the aura of the utilitarian constructions in the lanes. This is conceptual art more than ecological slide-show. This replica is framed, free-standing, out of context, and perhaps therefore a more suitable subject for reflection than the 'real' thing.

As if to prove the point, alongside it is a modelled version of an Atlantic woodland. These sodden labyrinths of wood, full of crook-backed trees and draped with mosses and lichens, are a special treasure of western Britain. One would have grown on the site of Eden in the Stone Age. Eden's repro is a feisty piece, strewn with mossy boulders and rotting logs, and planted up with wood-rushes and ferns and gladdon iris. I'm smitten by it, and find I'm pacing up and down the edges, peering for beetles and unbidden seedlings. There are a few oak-lings, grown from a collection of acorns nursed by a local school, but the designers have sensibly declined to transplant any old hunchbacks into their 200 square metres of forest legerdemain. Instead the idea of an ancient tree presence is suggested by Kate Munro's stark metal sculptures, fingered sprouts of recycled steel and lead, bent away from the prevailing wind. Their hammered and sheared surfaces sound an echo of the texture of weather-beaten bark.

I visit Kate in her workshop up among the Portakabins. She's dressed in a battered terracotta welder's jacket and goggles, and is putting together more plants with aspirations. I recall that the wife of Vulcan, the archetypal metal-worker, was Venus, goddess of beautiful things. Vulcan himself was a forger, a word which picked up its second meaning of 'fabrication' as far back as Chaucer. Kate has also helped to make a woven willow hut in the Myths and Folklore exhibit. It's in the shape of a huge onion, and the live willow shoots are already beginning to insinuate their way up the sides.

What these pieces – and there are many artworks round Eden – seem to represent is the idea of *becoming*, of the living processes that link ancient beginnings with unknown future forms. They have some kinship with old ideas of sympathetic magic, of encouraging growth or change by parading images of these processes. Few people today would seriously believe that forging metal trees will foster the growth of real ones. Yet they act as talismans here, small charms of encouragement.

The late arrival of plant images in art is a puzzle. From the Middle Ages onwards they became one of the most pervasive features of painting and decoration. They were the ornaments of the emblematic 'flowery mead' and of Books of Hours and Calendar pictures. They had rich symbolic significance throughout the religious art of the Renaissance. In the expansive mood of the seventeenth and eighteenth centuries paintings of plants became one of the links between science and art. At much the same time it became routine to set portraits of human affairs in a vegetated landscape. Today, the earth itself becomes a partner in the making of art. The gentle assemblies of leaves and twigs made by Andy Goldsworthy, for example, rejoin the natural cycles that produced them.

Yet before this comparatively recent period, plants were minor characters in art. Among all the artistically breathtaking and culturally complex images of animals painted by Palaeolithic people in the caves of southern Europe, there are only a handful of possible plants. An antler harpoon with a hint of carved leaf-buds. A rudimentary fern-like form painted on a stone from Parpallo, Spain, about 12,000 BCE.

Why were plants ignored? They were, after all, profoundly important to Stone Age peoples. Wild fruits and nuts were their mainstay when meat was scarce. Hallucinogenic plants like henbane may have been used by their shamans in the rituals associated with the animal paintings. And as hunters they would have been well aware that their food animals – bison, wild horses, aurochs – moved around according to the growth cycles of plants. Were plants too mundane, too bereft of spirit to figure in the intense visions in the caves? Were they seen as a fixture in the landscape, a reliable extension of the earth, not at all like the unbiddable, capricious beasts which might or might not appear in their season? (But then plants appeared and disappeared too.) Or was the empathy with animals that is so obvious in the paintings – and so hard to achieve with plants – an inextricable part of the huge advance in consciousness that made the paintings possible?

It is not until long into the era of agriculture and plant domestication that images of plants – 'capturings', we still call them – begin to appear on friezes and pots. In Egyptian art, already well populated with birds and animals, notional plants appear from about 2500 BCE. And in the Great Temple of Thutmose III at Karnak in Egypt, dating from 1500 BCE, there are limestone reliefs of some 275 wild plants brought back by the king from a campaign in Syria, drawn as dispassionately as any modern scientific diagrams. There are recognizable pictures of pomegranate, dragon arum and large cuckoo pint. The plants were part of the king's loot, 'all the plants that grow, all the goodly flowers that are in the Divine Land'. Three centuries later, another Egyptian artist painted a fully kitted-out farm on one of the walls of Sen-nedjem's tomb in Thebes. The estate is surrounded by irrigation channels, and is very tidy. In one part of the fields, Sen-nedjem and his wife (or perhaps their servants) pick tall bushels of flax. In an adjoining plot he cuts what look like ripe barley heads with a sickle. There is an orchard of date palms, heavy with fruit. It is a painting of human beings unquestionably in command of their landscape. Yet, positioned as it is in a nobleman's tomb, it is almost certainly allegorical, too. The farm is a symbol of human life on earth. Just as the crops are cut down and then shoot again (but only if planted by humans) so the king and his wife will live again after death, reincarnated by the gods.

Yet Stone Age people did make one half-step towards representing plants. They painted bees, alchemical distillers of the essence of flowers, and they painted honeycombs. Bees were humankind's first familiars, the first animals *used* rather than simply killed and eaten. And in a limestone rock shelter near Valencia in Spain, there is the first painting found in Europe which is recognizably of honey-hunting. The hunter is clinging precariously to a set of long ropes, dangling down over a rock face. One hand is reaching into a circular honeycomb; the other holds a bag with a handle. A dozen bees are arrowing in towards the robber, making a bee-line for him. They are huge, bigger than the hunter's head. Their wings are furious scribbles. The hunter has a long way to fall. Honey is worth it, the picture says.

And among the teeming horses and bison that thunder about the ceiling of the great cave of Altamira, near Santander in Spain, are strange geometric patterns which may also have connections with bees. One is a series of concentric curves, nested semicircles, likened by one observer to 'a child's drawing of a rainbow turned upside down'. Identical images occur at many rock-art sites in southern Africa. Those who've studied the history of honey-hunting have come up with a plausible explanation. A hunter who has just cut into a tree trunk and taken the front from a wild bees' nest would see the hexagonal chambers in the honeycomb as a series of parallel semicircular curves, with the largest in the centre. And, as in a honeycomb, the geometric shapes on the rocks are darkened at the centre (where the young bees are developing) and paler outside, where the clear honey lies.

Yet perhaps, argues David Lewis-Williams, most audacious and maverick of Palaeolithic art interpreters, the honey-hunter also glimpsed in the honeycomb an image identical to one he'd witnessed in a trance state. Glowing, geometric patterns – known as entopics – are commonly 'seen' by people in trance states, regardless of whether these are induced by drugs or dance. The glimpsing of a honeycomb at such a moment would enhance the status of the bee as a sacred insect, and honey as a magical food. The artwork became a homage, a confirmation of the inseparability of humans and nature.

The hexagons in the biome walls are partially reflective. Looking at them from the outside you glimpse clouds scudding across rainforest trees, the sun rising over orange blossom. Eden is all about legerdemain, about conjuring with plants. The whole Project is like a monumental work of cave art, a supplication to nature. Individual exhibits are both self-contained plant communities and representations of them. The translocated Cornish hedge and the oak wood mock-up must simultaneously perform as real places and as ciphers, parables about the world outside.

If they were actually out there in the 'real world', instead of in this theatre of ecology, we might, conceivably, be worried about their authenticity. This, in an increasingly virtual world, is a quality that may be losing its appeal. In Nevada, there is a mock-up of one of the most beautiful human artefacts on the earth, the centre of Venice. A lookalike Rialto Bridge arches over a four-lane freeway. In France, a faultless reproduction of the Lascaux cave-paintings has been built to safeguard the original. Visitors flock to both these facsimiles. Yet the stamp of the 'real thing' continues to exert a stubborn appeal. The knowledge that the original Lascaux bisons were smeared in by Palaeolithic fingers fifteen thousand years ago gives them an aura that can't be captured by any simulation. Authentic lineages are not something that can be recreated. An ancient forest, with a continuous history back to the Ice Age, is not the same as an identical collection of planted trees. It not only has irreplaceable ecological subtleties but a unique engraining of *time*.

The thoughts Walter Benjamin expressed in the 1930s in 'The Work of Art in the Age of Mechanical Reproduction' transfer plausibly to the relation between real and facsimile 'nature':

> Even the most perfect reproduction of a work of art is lacking in one element: its presence in time and space, its unique existence at the place where it happens to be. The unique existence of the work of art determined the history to which it was subject throughout the time of its existence. This includes the changes which it may have suffered in physical condition over the years as well as the various changes in its ownership ... One might subsume the eliminated element in the term 'aura' and go on to say: that which withers in the age of mechanical reproduction is the aura of the work of art. This is a symptomatic process whose significance points beyond the realm of art. One might generalise by saying: the technique of reproduction detaches the reproduced object from the domain of tradition.

For Eden we could reframe this as 'Nature in the Age of Mechanical Reproduction'. But we shouldn't push the similarities between nature and art too far. To view the natural world *as* a work of art, as a finished, achieved product (and, with increasing presumptuousness these days, as a product in some way fashioned by us), is one of the damaging conceits of industrial societies. Yet analogies reverberate through Benjamin's essay: the deteriorating physical condition of the 'originals' (in Eden's case, the biomes of the planet itself), the desire of a growing public for access to images of these; yet a hankering, too, for that aura of naturalness.

To find a way of fulfilling these often contradictory goals is the Eden Project's challenge. It is both a real place and a symbol, a web of communities that work in their own right, and narrate the working of others in the world beyond. Its aim, by its own account, is to educate and entertain, to use the idea of 'a living theatre of plants and people' to awaken visitors to the crucial role of plants on the planet. Yet slowly, inexorably, places like Eden are *becoming* the planet. The natural world is now increasingly contained, both physically and in our minds, in enclosed reserves and managed gardens, in simulations and virtual experiences. Paradise has become a fenced enclosure.

Thorns and thistles

INSIDE THE WARM TEMPERATE BIOME, I could be back in Crete. Pink flowers of Cretan cistus, with the texture of crimped paper, cover the bushes. Waxy leaves of sea-squill (whose bulbs the Cretans use as rat poison) flop among the imported boulders. The air is full of the scents of the south – resin, broom, orange blossom, dust. The biome has representations of the vegetation from several of the Mediterranean-type climate zones of the planet, including South Africa and the Californian coast. But it is the surrounds of Homer's 'wine-dark sea' that dominate the place. You begin your tour through the entrance to an Italian villa, complete with frescoes. 'The Mediterranean Basin. Culture's Cradle: Made by Hand'. The notices are unambiguous, both about the natural settings and their domestication: 'They have hot, dry summers and cool, wet winters. Their wonderful profusion of plants springs from a human gardener's nightmare of drought, poor thin soils, and fire.' But the 'natural Mediterranean vegetation has been cut for timber and firewood and cleared to plant crops for thousands of years. Without the olive and the vine, civilisation might never have begun. The "natural" landscape we see today is the product of both nature and humankind.'

I climb up the track between terraces of granite boulders. There are wispy sheaves of a lavender whose smell is not lavender but a rough, skirling mixture of hyssop and camphor, a baked hillside scent. (Not all botanists have felt the same about the scent. One likened it to 'burning rubber or a lion's house'.) I'd smelt and seen it once before, by a roadside in southern Spain in company with two other wild lavenders, fringed lavender, *Lavandula dentata*, and the silver-leaved, woolly lavender, *L. lanata*. This cut-leaved lavender, *L. multifida*, is a little gawky, with the flower spikes held on loose stems that in the wild don't often get above half a metre. In the reliable warmth of the biome they sprawl to two or three times this length. The feathered leaves are exquisite, like tiny ferns. One of the gardeners approves of my admiration. Cut-leaved lavender self-seeds here, he says, and grows so well that they have to thin it out every couple of years. But it is barely hardy. I wondered how I could grow it at home, without a greenhouse. John Gerard, ever on gardening's cutting edge, called it 'jagged sticados', and reported that 'we have them in our gardens, and keepe them with great diligence from the injurie of our cold clymate'. In the eighteenth century gardeners used to make little cloches out of oiled paper or sailcloth to house the tender plants pouring in from the four corners of the world. Or they protected them with inverted bell-jars, like miniature biomes.

All the visitors are at the same business, spotting favourite plants, sighing at new ones, amazed at how many English garden favourites originated in the sun-baked hills of the Mediterranean. I'm a subject of envy for having a notebook to write their names in. A lady from Holbeach taps me on the shoulder and says she hopes she can remember the name of the bright cerise geranium from South Africa that she's fallen in love with.

Eden draws in like-minded people just as any place of pilgrimage does. Today it feels like an old school reunion. I've already bumped into a clutch of old colleagues: Hugh Synge, editor of *Plant Talk*, Jane Smart, director of Plantlife, and Richard Sandbrook, first head of UK Friends of the Earth, whom I'd last seen when we sat on a committee together deciding what place would be best suited as a memorial to the writer and broadcaster Kenneth Allsop. (We opted for Steepholm island, the one site in Britain where the wild peony grows.) Typically, Jane and Richard were here to discuss forming a company to publish Hugh's magazine.

And then I bump into the entire family of the old friend with whom I made my first trip to the Mediterranean, when we were both in our early twenties. They hail me from the aromatic heights while I'm meandering through the lowland crops. Peter Newmark and I have often bird-watched together, and our talk shifts naturally to the robins that are darting about close to our heads and feet. They shuttle between plant-foraging and making little entreaties to the visitors – a shiver of wings and a brief gape of the beak that's an echo of their food-begging performances as fledglings.

These birds are features that weren't included in the designers' blueprints. But for the last two years they've been stealing in through the ventilation panels, and stalking visitors through the doors. Now they're breeding among the cistus, and in the morning the biome rings with their songs. But with so much ecological artifice on hand, Eden has felt obliged to present a certificate of provenance for the robins. 'The birdsong that you hear is real,' announces the notice propped up among the bushes, 'and belongs to our robins that have made the biome their home.' Later I saw blackbirds, wrens and chaffinches all feeding and seeming thoroughly at home. They are, I'm assured, quite able to find their own way out, but some looked thin, and listless. I find I'm unsettled by the way these wild birds look you in the eye and beg.

It's like being in a Mediterranean *passeggiata* here. You join the procession, chatting, peering at your fellow strollers, pointing at curiosities on the way. The cognoscenti tap pots and sift soil through their fingers. All that's missing is a bar. So I sit down instead by a lookout point that gives a view clear down over the cannily contrived terraces to the vineyard with its Bacchanalian sculptures. A sharp, dark-clothed, grey-haired man walks past me. He's striding purposefully, as if he knows the place. Ten yards on he stops, turns round and walks back towards me. I'd last seen Dominic Cole twenty years before, when he was a tyro landscape designer at a specialist consultancy. Now he is Eden's principal landscape architect. He was responsible for the overall feel of the biome, for the positioning of terraces and beds, for the broad strokes of the planting plan and the crucial finishing details. I'd known Dominic since he was a boy. We'd grown up in the same corner of the Chilterns, and his elder brothers were in the same teenage tribe as me. When he was small he dismayed his parents by devoting himself to flower-arranging instead of war games. Now he's the most accomplished flower-arranger on the planet. He comes back regularly to see how things are 'coming along', and even as we're discussing the problems of transplanting hundred-year-old olive trees, his eyes are darting about to check the state of the furniture and the disposition of the plants. He's not sure about the awnings above us, which look as if they're intended to suggest a stack of Ibizan deck chairs, but is pleased with the way his handrails ('modelled on the beadings in the Alhambra') are weathering.

The landscape of aromatic shrubs and strewn stones that dominates this section of the biome is known, the displays tell us, as *garrigue*. This is a French word, but this scrubland is such a dominant habitat throughout the Mediterranean region that it has a different name in every country. In Greece it's *phrygana*, in Palestine *batha*, in Spain *tomarilles* or, more often, named after its dominant species: *romaral* is rosemary scrub, *xaral* an area of cistus. It is a landscape of great symbolic significance, not just because of the romantic appeal it holds for travellers but because its diversity and resilience hold many lessons about the meaning of fertility in the natural world.

Garrigue is generally defined as a layer of undershrubs that reaches up to a metre or so in height, plus lower layers of spring bulbs and early summer annuals. It is one of the most colourful and diverse plant communities in the temperate world. Throughout April and May the washes of silvers and grey-greens in the foliage are lit up by the pink and white rose-flowers of cistus, by violet spikes of lavender, by the fat, yellow, lippy blooms of phlomis and bright tufts of thyme and dittany.

The first *garrigue* I ever saw was in the Languedoc in south-west France. It wasn't, as it happened, especially rich in species, but was composed – I can think of no other word – of such extraordinary and visually compatible plants that it was unforgettable. The groundbase was thyme, cushions of downy-leaved, frothy-flowered common thyme. Around them were other low-growing tufts: yellow flax, and *Aphyllanthes monspeliensis*, an evergreen member of the lily family. Its bright blue flowers were held at the very end of its rush-like stalks, which spilt out like a posy whose ribbon has been cut. Of straightforward vertical plants, there were just two: white asphodels a metre tall, and the low, runic flower-heads of spider orchids.

The richest *phrygana* I have seen was in Crete. The shrubs were a kaleidoscopic improvisation in yellows, a mosaic of asphodel, Jerusalem sage (*Phlomis fruticosa*), chrome-flowered and coconut-scented thorny broom (hard to tell from the furze that was in full flower back in Britain) and mounds of lemon-green Greek spiny spurge. As the sun set, the scrubland glowed as if it had been burnished. In the more open grasslands between and above the shrubs we found, in one morning, twenty-four species of orchid, including the monkey, tongue, and naked man – *Orchis italica* in Latin and *Uomo nudo* in modern Italian (and with a 'central appendage present', as the guide books politely put it, in the manikin-shaped flowers). There were Cretan tulips and sheaves of wild irises, including the widow iris, *Hermodactylus tuberosus*, and the delectable Barbary nut, *Gynandriris sisyrinchium*, which opens so smartly (at about 2 p.m.) that you can actually sit down and watch the pale blue petals unfurling like sea anemones. 'So we have the mourning iris,' pronounced my companion Bob Gibbons, in one of his best botanical puns, 'and the afternoon iris.'

But what is often more striking about *garrigue* plants than their looks is their scent. The majority of the shrubs – rosemary, helichrysum, lavenders and sages, the gummy cistuses – are sharply aromatic. To walk through a *garrigue* is to be brushed and combed with sticky, resinous foliage, and made heady by the mixture of balsams and volatile oils. This isn't an accident. The thin coating of oil-bearing cells that so many of the plants have on their leaves serves two functions. The bitter oils provide a deterrent to browsing animals and a protection against the sun, a way of conserving precious water resources. As well as acting literally as an oilskin, the oils evaporate in the heat and keep the plant cool. (Other devices employed by leaves for the conservation of water include coats of wax or closely packed hairs, or rolling up like tubes in fierce heat.)

There is a price to pay for such high concentrations of oils, though, and that is high flammability. *Garrigues* catch fire (or are set on fire) with regularity. It's a natural phenomenon in such hot, dry regions, and just as the *garrigue* species are adapted to resist the heat, so they have also evolved to regenerate quickly after even quite bad conflagrations. Several kinds of cistus, for example, produce two sorts of seed: a soft-coated one that, if there is no fire, germinates simply by being wetted, and a hard-coated one that will germinate only after being heated. Fire seems positively to benefit many species. Underground bulbs are stimulated to flower – sometimes only days after the fire. During the following winter the burned area becomes covered with young plants, thriving in the increased light and moisture reaching the ground, and from the minerals recycled from the ashes.

This adaptation to fire is also true of *garrigue*'s larger relative, known as maquis. Maquis (which lent its name to the French resistance fighters of the Second World War, who hid out in its fastnesses) is a layer of shrubs or small trees up to five metres high. It consists, typically, of evergreen shrubs like juniper, box, strawberry tree, stone pine, wild olive, mastic (*Pistacia lentiscus*, whose resin is used as a chewing gum) and sumach, which form a closed canopy – recognizably a 'shrubbery'. Some of these species can grow quite tall, and maquis easily develops into woodland. After fire or heavy grazing it can also turn into – or be invaded by – *garrigue*. The reverse process happens less easily. Established *garrigue* is usually on the thinnest and driest soils, and consists of shrubs that can't grow to great heights, and which aren't quickly invaded by larger tree species. But there is no clear dividing line. Juniper and cistus, for example, can live in both vegetation types.

The species that most clearly demonstrates the versatility and anciently evolved resilience of the Mediterranean's plants to its harsh conditions is the prickly or kermes oak, *Quercus coccifera*. Prickly oak grows throughout the region, though it has a slight preference for limestone soils. It's a gregarious, evergreen, acorn-bearing oak with usually rather tough, spiny leaves, but beyond that it's so variable that it's hard to give a simple description. In very poor soils, or under pressure from grazing or fire, it can prosper as a 'ground-oak', never reaching more than a few centimetres in height. At the other extreme it can make a twenty-metre-tall, three-metre-thick tree. Trees of this size in the Lassithi mountains in Crete are over a thousand years old.

Prickly oak can pass effortlessly between these two states. It isn't killed by burning, felling, or browsing by goats and sheep. It simply regrows from its stool or rootstock, and is very long-lived. Nibbling by goats tends to give the regenerating oaklet a columnar shape. The shoots at the circumference of the bush spread outwards until the browsing animal is unable to reach the centre shoots – whereupon the oak 'gets away'. It may eventually become a full-sized tree with a low, barish trunk – at which point goats can get at it again, scrambling up into the branches and devouring the foliage as they do on the ground. The resulting 'goat-pollards' are highly distinctive, with sheaves of browsed-bare branches shooting vertically from the main horizontal ones. After fire, prickly oak can grow a metre in a year, and begin to produce acorns in its first or second year.

There is a third alternative to mature closed woodland, important because it is so widespread in the Mediterranean. Savanna is open woodland with patches of grassland, *garrigue* or maquis alternating with groves of taller trees.

The Mediterranean shrublands are rich habitats for animals – birds and insects especially. They've been diverse and productive environments for humans as well. They've supported herb-gatherers, asparagus-pickers, boar-hunters, bee-keepers. Topiarists, too. In Provence I once saw a patch of *garrigue* where the wild box had been clipped into fantastic shapes by a crew of telephone engineers with time on their hands over lunch.

In Crete, much of the population – from cities as well as the countryside – rummages about in the *phrygana* in the spring. On fine weekends in March and April it is an old social tradition to gather *stamnagathi*, edible wild greens. Families spill out of cars and, using little mattocks, dig up plants by the root – especially the young rosettes of spiny chicory, which are eaten whole or boiled in salads and soups. Both the ritual picking and the chicory's bracingly astringent taste are seen as signs of spring, and a challenge to the torpor of winter. Smallholders gather the bronze-red wild tulip, *Tulipa doerfleri*, which grows like a weed in their arable plots. In an area of eastern Crete, ladanum, one of the ingredients of incense, is gathered from the leaves of *Cistus creticus*. The technique, and the tool used, are exactly the same as recorded by the French traveller de Tournefort three centuries ago: 'a sort of Whip with a long handle, with two rows of straps ... the Straps whereof, by rubbing against the Leaves of this Shrub, lick'd up a sort of odoriferous Glue sticking on the Leaves ... When the whips are sufficiently laden with this Grease, they take a Knife, and scrape it clean off the Straps, and make it up into a Mass or Cakes ..'

Perhaps the oddest of these foragers' economies is visible on the oldest plants on Crete, the great cypress trees – some of them believed to be more than four thousand years old – which grow in the gorges and on the edges of the hillside *phryganas*. Some of them have what seem to be oblong scars, of the kind that occur naturally when a branch is lost. But their edges are too regular, and in fact they are marks caused by the deliberate cutting of rectangles of bark to make the lids of beehives. The cypress trees are unconcerned about this modest harvest, and continue to grow, their roots drilling down into the hillsides like seams of rock.

Throughout the Mediterranean, wild lavenders are perhaps the most used of all the *garrigue* herbs. There are a score of species and sub-species in the region, and most have been employed in perfumery, cookery or medicine. Even cut-leaved lavender, with its pungent 'lions' house' scent, is used in Morocco in infusions against coughs. In France, the Provence town of Grasse has been a focus of lavender processing since the Middle Ages. During the Renaissance, under the influence of the Medici family, it developed into the commercial centre of the perfume industry. While distillation factories were established in the town, the surrounding countryside was used for the production of the raw materials – not just the growing of jasmine, roses and irises, but the raising of sheep, which supplied leather for the tanneries (for the making of scented gloves) and the fats used in soap and cosmetic creams.

At the close of the nineteenth century, the harvesting of lavender underwent a long process of transformation that might be a parable for the social and economic changes that happen during the switch from foraging to large-scale cultivation. During the early 1800s, the rural areas of southern France suffered massive depopulation, as a consequence of industrial development. There was an exodus of peasants to the cities, and the abandoned farms were quickly colonized by wild lavender, *Lavandula angustifolia* – especially the variety (subspecies *delphinensis*) known as *lavande moyenne*, which thrives in stony, previously worked soil. At the same moment, the demand for perfume from the growing population in the towns began to escalate, and those small farmers left in the countryside realized they had a valuable raw material at their disposal.

To begin with, gathering the wild lavender was a job for women and children, or shepherds in their spare time. The flower spikes were brought back in canvas bags, or tucked into aprons. In some places the pickings were measured in 'aprons'. By the 1850s, the operations had become more organized. The heads of families took charge of the teams of pickers, which included not only family members and neighbours but seasonal and itinerant workers from the gypsy community and from Spain and north Africa. The teams could number up to thirty or forty, and the production of lavender oil was carried out with portable stills, taken up by mule to the nearest water source. But soon the economies of scale crept in. Permanent distilleries were built on fixed sites. The professional distiller and oil-broker arrived, buying the harvest from the pickers and selling the essence on to big manufacturers.

Recognizing the value of the wild crop, the farmers and pickers then began a kind of proto-cultivation in an attempt to increase its yield. They thinned plants, grubbed out shade-bearing species like pine, and sometimes applied manure (though this could increase leaf growth at the expense of oil production in the plants). They tried transplanting young wild lavender to cultivated fields, but this was rarely successful, especially without water for the traumatized root systems.

Large-scale cultivation didn't really develop until the 1920s, and then not with lavender itself but with the hybrid form, lavandin. *L. angustifolia* is very much a hill species in Provence, growing at 500 to 1,500 metres on limestone rocks. Spike lavender, *L. latifolia*, tends to occur lower down, at between 200 and 800 metres. The hybrid, *Lavandula x intermedia*, known as lavandin or *grosse lavande*, appears in the zone where these two species overlap. The hybrids stand out in the landscape because they are more robust, more vigorous and longer-lived than their parents, and experienced pickers soon began to select them. They proved to be hardier and more adaptable, too, and to have a yield of oil which could be ten times that of ordinary lavender.

Lavandin is sterile, and propagation could be achieved only by taking cuttings. These were reared in nurseries, and transferred to open fields after eighteen months. The creation of these clones proved easy, and lavandin cultivation became widespread during the 1930s, not only in the small field systems of the hills but in the richer, potentially arable zones in the valleys. But like all clones, they were easy targets for disease, and a severe yellowing ailment (*dépérissement*) became endemic in Provence, reducing the lifespan of the plants from ten years to three or four. Resistant strains were bred in the 1950s and 1960s, though the disease still reappears. The gathering of wild lavender as a commercial business ceased in the 1950s.

In 1923 100 tons of lavender oil was produced, 90 per cent of it from wild plants. The cultivated forms accounted for the remaining 10 per cent. Today, *L. angustifolia* accounts for just 25 tons of distilled oils a year, none of it from wild plants, while the quota from cultivated lavandin is more than 1,000 tons. En route, lavender farming has acquired all the characteristics of an industrial monoculture. The dense rows of plants are sprayed with herbicide twice a year, harvested exclusively by machine, and have to be intercropped with soya beans or sunflowers to combat their increasing susceptibility to parasites and diseases. Ironically, the commercial products that keep this now large industry in business are synthetic detergents, for which lavender essence is the scent of choice.

Paradise lost?

THESE ALL SEEM BENIGN HUMAN PICKINGS from a rich and varied environment. Yet there has been a persistent belief that the Mediterranean is an irretrievably degraded landscape, a one-time sylvan paradise which has tumbled down into a desert – and that human use is the culprit. The great historical ecologist Oliver Rackham calls it the theory of the 'Ruined Landscape'.

One of its earliest surfacings was in Plato's description of the deforestation of Attica in 400 BCE, though what hard evidence he had of the region's previous state is unknown: 'What now remains compared with what then existed is like the skeleton of a sick man, all the fat and soft earth having wasted away, and only the bare framework of the land being left.' In Dante's *Inferno*, written in the fourteenth century,

there is a description of 'a waste land ... which is called Crete, under whose king the world was once innocent. A mountain is there which once was happy with water and leaves, which is called Ida; now it is a desert like an obsolete thing.' In the seventeenth century painters like Nicholas Poussin bolstered the story by setting their studies of Greek myths in idyllic woodland landscapes. Poussin lived most of his life in Rome, but never saw other parts of the Mediterranean. Like many Baroque painters and Renaissance poets, he encouraged a belief that the great dramas of classical times were acted out in landscapes not unlike the lush woodland of central Italy. When travellers 'discovered' the more arid parts of the Mediterranean region and compared them with their expectations, they concluded that something had gone badly wrong.

The theory lost none of its assertiveness in the more scientific twentieth century. The distinguished archaeologist Sir Arthur Evans stated that the builders of the temple of Knossos, in Crete, turned to the use of gypsum for door and window frames because, as early as the Bronze Age, they had run out of trees! In essence the Ruined Landscape theory sees the scrubland and savanna habitats of the Mediterranean as no more than the crumbling relics of a landscape which is played out, culturally and ecologically. In 1985 a major BBC television series about the Mediterranean made a succinct elision between these two kinds of Fall. It was called *The First Eden*, a reference not to the area's present condition but to its past. Generalizing from the effects of the Venetian Empire's naval adventures, it suggested that by the end of the sixteenth century shipbuilding in Venice had ceased because 'the Mediterranean forests were now almost exhausted of timber.' After the trees had been destroyed, the fate of the stricken land was sealed by the goat (an animal which had been around in the Mediterranean for four thousand years). Goats, unlike sheep, don't demand grass and will eat anything, however fibrous or thorny. And once goats are established, 'the land stands little chance of recovering its trees and regenerating its topsoil. The goats consume every seedling that sprouts and every leaf that unfurls.' The forests that once ringed the entire Mediterranean were largely gone. 'Now the maquis and *garrigue*, which had once grown only on the rockiest and most impoverished stretches of the coasts, spread all around the sea and dominated what were once among the most fertile lands in the world.' (Interestingly, this idea of a Fall in the landscape is never offered as an explanation for the other mediterraneoid regions of the earth, in California, south-west Australia and South Africa, where the equivalents of *garrigue*, such as chaparral in California, are even more widespread.)

Does it matter whether this account of the Mediterranean's history is strictly true? It does after all sound a general warning against the over-exploitation of nature, and that can be no bad thing. Yet its roots may be neither as innocent nor as well-meaning as that. The theory of the Ruined Landscape has, too often to be a coincidence, been an argument of convenience for those with a political agenda. It is used by commercial foresters who would like their plantations (often of foreign trees like the eucalyptus) to be seen as 'replacements' for the Lost Forest. It's employed by landowners and large-scale farmers who wish to denigrate the 'inefficient' techniques of peasants and smallholders. It has, en route, overstressed the importance of the 'fertility' of the soil, and denied the importance of fire as a natural agent of regeneration. As, at best, a generalization for such a huge area as the whole Mediterranean, it ignores the intricacy and detail of both cultural and natural systems.

And it continues to give support to another corrosive legend prevalent in the developed world, that between pristine wilderness and intensive agriculture there is no viable middle way. In this view of the living world, humans are still irrevocably exiled from Eden.

The theory of the Ruined Landscape isn't a myth. It has no ancient roots in the human imagination or any existing culture. It does nothing to help us understand our place in nature. Oliver Rackham calls semi-fictional inventions of this kind 'pseudo-histories'. They are the province of received wisdoms and convenient maxims, and are politically and ecologically dangerous.

In among the lavender and the sages, I spot a small notice with a quote from the French historian Fernand Braudel. It's given without comment: 'When you don't cultivate land in the Mediterranean, the land dies'. What does he mean? That before the arrival of agricultural man, the Mediterranean was a true desert, devoid of life of any significance? That it was a waste – or wasted – land? That now its only liveliness is the consequence of what human 'cultivators' put into it? At first sight it seems a view directly opposed to the Fallen Paradise theory. Yet both share an implicit assumption that humankind and nature cannot be equal partners in the landscape, that it is only from 'improvement' by man that nature realizes its full potential.

Just a few yards on from this quote the pathway passes a rosemary hedge. It's about six feet long, and waist-high. Almost everyone who notices it as they walk past runs a hand along it, strokes a branch, pinches a shoot, raises fingers for a sniff. The hedge has been worn smooth by this affectionate petting. But only on one side. On the unreachable far edges, the rosemary grows out shaggily above the terraced beds below. Is this hedge a metaphor for the elusive borderland between cultivation and the wild? The same plant, free-range and ragged on one side but available and controlled, however accidentally, on the other?

I first encountered Oliver Rackham back in 1974, and the circumstances are worth noting, given what he had to say. He was speaking at a major symposium on 'The English Oak' at the University of Sussex, which had been partly planned as a covert undermining of foresters' view of the oak as a kind of arable tree, utterly dependent on 'good seed' and human sowing. He had no book to his name then, and was unknown outside specialized ecological circles. He was a modest speaker, not given to theatrical declamations. But what he had to say was, in the context of the conventional forestry wisdom of the '70s, a revelation. The general direction of his argument – the talk was called 'The oak tree in historic times' – wasn't entirely unfamiliar. He spoke of continuous yield systems which had been practised in woods since prehistoric times, especially the cropping of fuel and rough wood by coppicing, and about the harvesting of young maiden oaks for house-building. 'Contrary to popular belief, the harvesting of woodland produce did not destroy the wood. When a man felled a tree, he expected it to be replaced, normally without artificial replanting.'

But it was the detail and texture of his supporting evidence that riveted the large and sophisticated audience. He shunned forestry

maxims, unquestioned legends, unsubstantiated evidence from subjective and anecdotal sources – all the kinds of material that had bolstered
the conventional wisdom of trees since at least the eighteenth century.
He had gone instead to first-hand evidence, in woods themselves, in
the timber in buildings, in the unromanticized details of bills of sale
and carpenters' accounts. He had counted and measured all the tim-
bers used in the original building of his Cambridge college, Corpus
Christi (there were 1,249). The vast majority were not 'mighty oaks', but
squared trees between six and a half and eight and a half inches in
diameter. Very few were longer than twenty feet. The age at which they
were felled, confirmed by ring-counts where possible, was between
twenty-five and seventy years. He had also counted the smaller
number of trees in a modest Suffolk farmhouse from the period of East
Anglia's Great Rebuild (c. 1600). It had used seventy-two oaks of
between six and a half and twelve and a half inches in diameter.
Knowing the stocking rate of young oak trees in Suffolk (twelve to
forty an acre) and having measured the rate at which they could grow
(seven feet a year from promoted coppice shoots in Bradfield Woods),
Rackham estimated that this represented the annual oak increment on
about seventy acres of woodland, and that the average parish could
have produced one such house every six months, if there were no
other demands on its timber. There weren't often such demands.
Neither shipyard nor woodland records show more than occasional
consignments of East Anglian oak for shipbuilding, against the popular legend that Britain's oakwoods were devastated for the navy. This
is almost certainly another piece of pseudo-history, concocted for
political and economic reasons, and is easily refuted by a look at the
rate of growth of the British merchant and naval fleets. The graph rises

slowly and steadily during the 1600s and 1700s, and only takes off sharply at the very end of the eighteenth century, when the merchant fleet increased eightfold to four million tons in little more than fifty years. (The navy showed a slight and temporary boom during the Napoleonic wars.) By this time, quite different factors were affecting the availability of timber, including the loss of woodland because of Enclosure and agricultural improvement, and the beginning of large-scale imports of naval timber from Albania after 1809. Naval timber was cheap, and the price remained unchanged throughout the second half of the eighteenth century – hardly an indication of shortage. As Rackham writes: 'Had there been the slightest physical difficulty in finding timber for the tiny fleet that defeated the Armada, it would have been utterly impossible to build the sixty-fold larger fleet that defeated Napoleon.' (The legend of the dire effects of shipbuilding on the nation's timber supplies is often linked to another plausible but equally groundless belief, that the beams in timber-framed houses were recycled from ships. The whole story comes close to the kind of symmetry that one associates with real myths: the hearts of oak were cut down, but it was in a good cause, and in the end the old salts returned to the hearth.)

The details here are worth elaborating for two reasons: first, because they shed light on some of the beliefs about the impact of shipbuilding on Mediterranean woodland; and second, because they are the substance of Rackham's philosophy, and as far as it's possible to be from trivialization. Detail is a measure of the intricacy and diversity of nature (and the inventiveness of humans). It's a register of the character of place, and the source of nature's resilience.

Twenty-five years later, and after a clutch of books that changed the face of the historical ecology of Britain, Rackham turned his attention to southern Europe. With his Cambridge colleague, the geographer A. T. Grove, he published in 2003 *The Nature of Mediterranean Europe: An Ecological History*, which warns against facile generalizations about such a vast and diverse region. In a wonderful litany that links the details of place, nature and work, he puts on record that the Mediterranean includes

> sugar-boilers in Motril, acorn-eaters in Estremadura, pig-driers in the Alpujarra, esparto-twisters in SE Spain, palmists in the city of Elx, madder-growers in Provence, cork-cutters in Sardinia, boar-hunters and chestnut-millers in the Apennines, oat-growers on the Macedonian serpentine, cotton-pickers in Boeotia and (formerly) Crete, resin-tappers in Attica, quail-gatherers in the Mani, banana-men in Arvi, ladanum-whippers in one particular spot in north Crete, potatoists on the Lassithi Plain, distillers of lemon leaves in the Cyclades, shipbuilders in the remote nooks of the Aegean, masticators in the southern third of Chios, spongers on the Twelve Islands.

Another register of the diversity of the Mediterranean is the flora itself. There are at least ten thousand species of flowering plant growing wild in the region, many of which occur nowhere else. The list is dominated by species adapted to drought, and to growing in hot sun: evergreens, summer-dormant bulbs, annuals. This becomes even more apparent when you look at the character of the most highly adapted Mediterranean plants – the endemics, species which are confined to one more or less isolated habitat, such as an island, a gorge, a cliff.

Crete has some 250 endemics, of which only twenty-one normally grow in the shade, and only eight could be classed as woodland plants. This is what would be expected where plants have evolved among a variety of open habitats, and casts doubt on the theory that the aboriginal Mediterranean landscape was unbroken forest from seashore to high mountains.

The evidence about the nature of the Mediterranean landscape given by ancient pollen remains in peat also suggests great diversity. But it has to be interpreted carefully. A wood of wind-pollinated trees produces far more pollen than the same area of insect-pollinated bulbs, so the bulbs' pollen-grains 'count for much more' than the trees'. Rackham has devised what he calls a Non-Forest Index (NFI) to adjust for this. 'As a rough criterion we propose to infer a substantial phrygana element in the landscape if the total undershrub pollen exceeds 1 per cent of the tree pollen. The full formula is NFI = Undershrub pollen + one-tenth of wind-pollinated non-shade-bearing herb pollen + insect-pollinated non-shade-bearing pollen / total tree-and-shrub pollen.' (This is rather numbing arithmetic. But for those with a taste for it, the NFI at a range of Mediterranean archaeological sites suggests a huge diversity of habitats in pre-farming days: continuous evergreen forest in Padul, Spain (NFI 0.7), savannas of deciduous oak with some *phrygana* around Lake Trikhonis in western Greece (NFI 5.5), patchy oak and beech forest with abundant undershrubs and herbs round Lago di Martignano in Italy (NFI 10.4) and pure *phrygana* and maquis at the windswept coastal site of Tersana in Crete (NFI 173).)

The balance of forest and *phrygana* in these mosaics would change according to shifts in climate and grazing pressure. A few centuries of damp weather would increase the tree component. Wild-fires at a time of drought would expand the area of shrubs, as would a temporary increase in the populations of browsing animals. (And there were plenty of these in pre-agricultural times, including, on some of the islands, dwarf hippopotamuses.) As we've seen, the Mediterranean flora is anciently adapted to both fire and browsing.

Of course the arrival of agriculture made dramatic changes in the landscape. There was large-scale woodland clearance for cultivation and a rise in grazing and browsing pressure. But there is little solid evidence that these new developments irrevocably transformed the character of the Mediterranean's landscapes – and certainly not that it was the cause of some irreversible degradation. For many periods it is hard to disentangle those alterations brought about by climatic changes and those introduced by cultivation. In some ways the landscape changes brought about by indigenous agricultures fitted into the natural swings between grazing land, shrub and wood. Some practices may even have generated a kind of symbiosis, beneficial to both humans and plants. The tradition of transhumance, for instance, the taking of sheep from the lowlands up to the mountain pastures in spring, gave animals more opportunity to forage in and exploit different habitats than if they had been confined to small, fenced enclosures. And it benefited the landscape by producing a variety of niches where non-intensive grazing and browsing had occurred.

The recent changes in the Mediterranean – the ribbon development of the coastal strip, the sweeping away of local farming traditions by the Common Agricultural Policy, the massive road-building programmes – have had an unequivocally damaging effect. Yet generalizations about the present are no more reliable than they were about the medieval Mediterranean. There are enclaves throughout the

Mediterranean region where both the landscape and the social and economic practices attached to it continue to be robust. Many would still seem familiar to observers from Plato to Shelley. And throughout almost the entire region, contrary to the core belief of the Ruined Landscape theory, native woodland is on the increase as peasants abandon the countryside. This may not be a universally welcome trend, and certainly doesn't indicate that the region is becoming a 'Sylvan Paradise' (again!). But it does suggest that in large and complex regions like the Mediterranean, it does not do to simplify the relationships between natural and human pressures.

As for that hangdog belief in the devastating effects of shipbuilding, beloved of Ruined Landscape acolytes both in Britain and around the Mediterranean, perhaps Rackham should be given the penultimate word, as he pits his scalpel against the bludgeons of pseudo-history. His account of modern Turkish timber-shipbuilding is a masterly demonstration of the triumph of living detail over second-hand generalization:

> Ships are not built in docks, but on any convenient piece of
> ground; they leave almost no archaeological trace. A typical ship,
> 25–31 metres long, is built of pine (*Pinus brutia*) with some elm and
> mulberry; other Turkish shipwrights use oak. The ship's frames are

made from short, crooked, savanna-type pines, some fifty years old and 50cm in diameter. Each frame consists of about twelve parts bolted together: a typical pine-tree produces about eighteen of these parts. The forty-five or so frames of a ship thus represent some thirty trees, which would have grown on about 0.4 ha of savanna. If we multiply this figure by three to allow for decks, planks and other parts of the ship, each ship represents about 1.2 ha of pines at fifty years' growth. The industry around Bozburun [SW Turkey] turns out some thirty ships a year, i.e. 1,500 ships in fifty years, and would thus be in equilibrium with the growth of pines on about 1,800 ha or 18 sq km of pine savanna. This is much less than the actual area of pineries on the Bozburun peninsula. If this is representative, it does not justify scholars' obsession with shipbuilding as the great consumer of trees.

The harvesting of self-renewing timber trees for house- or ship-building is one of the best examples of sustainable resource use. But it's salutary to remember that an environment of fifty-year-old oaks or pines is only sustainable from the point of view of the humans who use it and the organisms that have been able to adapt to it. It is no longer capable of sustaining those organisms (some of them human) whose need was for the fully grown and dead trees that preceded it. As I discuss later, 'sustainability' is a pliable word, and, ironically, is customarily used about limited slices of time and far from inclusive constituencies. This is a minor worry when talking about renewable wood harvesting, but crucial when we come to evaluate the impact of agricultural systems, and how they compare, ecologically and socially, with the landscapes they replaced.

The staff of life:
a rather flat stick?

THE PROCESSES THAT WERE RESPONSIBLE FOR the biggest changes in the Mediterranean landscape began in the extreme east of the region. The spread of cereal farming from Mesopotamia was one of the most significant changes in the history of humankind, but also one of the oddest. Why did crops and techniques specifically suited to the deserts of the Middle East come to dominate not just the more benign, wood-land-growing regions of Europe but every kind of potentially fertile area, from Chinese wetlands to clearings hacked out of virgin South American rainforest? What is it about the benefits of cereals that has consistently outweighed their costs in terms of devastated landscapes and social dependency?

What attracted the first farmers to the wild *Triticum* grasses that grow in the dry areas of the eastern Mediterranean is obvious. They responded to irrigation, had a natural tendency to grow in large clumps, and their seed-heads were easy to cut and harvest. The grains stored well, didn't rot in the arid climate, and, with rough grinding, could be made into convenient, roastable cakes. Yet the development of leavened bread from these early beginnings represents one of the most striking and curious transformations in human culture. Compared to the cooking of meat or pulses, say, the making of bread is astonishingly elaborate: the Staff of Life, across much of the planet, is the ultimate processed food. After the grain is prepared it goes through no fewer than four distinct transformations: milling, dough-making, leavening and baking.

Bread emerged very early. What amounts to the earliest cookery book in the world – discovered carved on a group of cuneiform tablets dating from 1600 BCE – shows that a bread culture was well established in Mesopotamia by the second millennium BCE. The cuisine employed at least four different grades of flour: coarse, fine, a semolina (the hard grain left after the sifting of flour) and groats (the hulled and maybe roughly crushed grains). Unleavened bread was made in a clay hearth, by sticking flattened doughballs on to the hot walls or, more elaborately, by mixing the dough with milk, oil and herbs, and baking in hot ceramic dishes. Leavened bread was being made a very long time before the discovery of yeast. The rising agent was sometimes a meat broth allowed to go sour, but most often beer. Predictably, for a culture in which grain crops were central, brewing was developed before the making of wine. The ideogram for beer, which appears in the earliest cuneiform writing from around the end of the fourth millennium BCE, represents a large vat filled with water and grain.

The virtues of a food economy based on grain are obvious. The physical nature of the crop – growing at a fairly even height above the ground – meant there could be economies of scale in its cultivation and harvest, and eventually mechanization of the entire process. The majority of grain crops are fairly nutritious, and, baked into bread, acceptable to most tastes. And grain-based agriculture enabled a certain type of society to develop. Arable farming is based on a narrow, specialized use of the land, and a similarly specialized use of human labour. This was reflected in the societies that mushroomed around it, in which the specialist jobs associated with cereal farming – accountancy, merchanting, farm management, tool manufacture, baking – prospered, and those that weren't connected had the freedom of time

to flourish. In this sense, cultivated cereals have been one of the linch-
pins in the development of urban and industrial societies. The catch is
that they are socially and economically addictive. Once a society is
organized around the ready availability of cereals, it isn't easy to see
how it could break the habit.

Is it partly to legitimize this dependence that the growing of corn
and the baking of bread are often regarded as almost sacramental
processes? It is hard to believe that cereals' rise to worldwide domi-
nance as our species' most important foodstuff happened without
being driven by, and forming, its own myths. There is a symmetry
about cereal-growing and bread-making. Agriculture – at least for most
of its history – has been the back-breaking grind in 'the sweat of thy
face' that Genesis promised. Bread-making, by contrast, is almost wilfully elaborate, a ritual that partly redeems the toil that led up to it. In
societies where foraged nuts and cultivated grains are both staples, the
wild harvest is invariably looked on as a sign of primitivism, a low-
status food that aspirant workers strive to rise beyond.

The ascendancy of another grass, rice, as the principal staple of
much of Asia probably didn't occur independently. The beginning of
rice cultivation is so late, comparatively (in the third millennium BCE in
India), that it was almost certainly inspired by the wheat-growing technologies of migrant farmers from the Middle East. But corn (or maize)
in the New World was taken into cultivation as a parallel, independent
process. No evidence of Old World grain crops was found when the
first settlers arrived in the Americas, and recently archaeologists have
discovered pod-corn cobs in cave settlements dated at about 3500 BCE
in New Mexico. Pod-corn is the wild ancestor of maize, yet the difference between the two types is fundamental: the seed-pod of wild corn

is covered by a closed sheath, whereas cultivated corn has 'naked' seeds. No such free-seeded variety exists in nature, yet it occurs as a frequent short-lived mutant in populations of pod-corn. The mutant is too vulnerable to predators to establish itself in the wild. But with the relative protection of cultivation it can survive, and it has obvious advantages to farmers in not having to be threshed free of chaff. This, then, was the origin of cultivated corn: an open-seeded pod-corn in central America – or just possibly the lowlands of south-west Brazil and Paraguay.

A number of South American tribal myths about the origins of maize have an uncanny resemblance to the Garden of Eden fable, even down to the central role of a tree. A typical story tells of the time when humans knew nothing about agriculture, and lived on a vegetarian diet of leaves and fungi. Then the existence of maize was revealed to them

by a woman who appeared in the shape of an opossum. The maize looked like a tree, and grew wild in the forest. But instead of harvesting the seeds, the humans cut the tree down, and then found that the dead remains wouldn't provide for their needs. They had no choice but to share out the seeds, clear the ground for cultivation and sow the first crops.

The myth fulfils two functions: it explains the origin of the different varieties of cultivated plant in the dispersed seeds of the maize; and it provides the model for a kind of Fall, in which greed, or a lack of wisdom and respect, is punished by the necessity of labour – which, in its turn, leads to salvation. As Lévi-Strauss has suggested, myths are attempts to resolve contradictions.

The Bed and Breakfast I was staying in housed a kind of arable myth of its own. Ivan and Kath threw up their jobs in the Midlands and came down to Cornwall to make a new beginning. Years before the Eden Project had been thought of they were recycling a ruined gardener's cottage on a big estate, and learning timber-framing and granite-chiselling on the job. Under the foundations of the building they discovered a tunnel, a securely built passage walled with granite boulders, and roofed with clean-cut stone lintels. It's a fogou, a structure built by Iron Age Celts. Cornwall is riddled with them, but no one is sure what they were for. They may have been used for storing grain, but some contain the remains of hearths and food fragments, and they may also have been lived in. Whatever their specific purpose, they are redolent of the adaptations that Mediterranean farmers had to make when they brought their skills north. Cereal crops, like soil, had always been a bank against the future. But what was their resilience, a thousand miles north of the Mediterranean climate in which they originated? These tunnels, sheltered against the moodiness of the Atlantic weather, have the feel of deep deposit boxes, safe havens for the crops under the earth that it was Adam's curse to till.

The Blazey fogou runs roughly south-west/north-east. An interesting modern ley line could be made by joining it with the Iron Age barrow at Prideaux and the corner of the Apple One car park at Eden.

The great drawback of exclusively arable systems is that they are two-dimensional. They reduce three-dimensional landscapes to flat drawing-boards, drastically simplifying their ecologies and social meanings. They are wholly managed systems, allowing little space for natural inventiveness or human ingenuity. They are single-minded and single-purposed, contrary to the rules by which living systems normally work. And this reduction, this homogenization, is reflected in the human societies that develop around them. No wonder that arable farming has so often grated with societies operating according to the principles of diversity and opportunism. The legendary conflict between the rancher and the farmer in the American West was an instance of this, even though the ranchers were hardly hunter-gatherers.

But are there viable alternatives for feeding large populations? Even in the Mediterranean there have been societies based on tree-products as staples. In the northern Apennines there is a whole culture (though it's run-down now) built around chestnuts. The trees are grown in terraced orchards, and large-nutted varieties grafted on to the best specimens. The nuts are dried over slow fires before being ground into a flour, which is superior in its taste and nutrients to wheat flour. Chestnut timber is used for beams in house-building. The grassland under the trees is grazed, burned occasionally to maintain its quality, and supports wonderful carpets of spring flowers. Other chestnut trees are coppiced on a large scale for firewood and vine-stakes.

Nuts can't be harvested mechanically like grains, so they are a labour-intensive crop – an advantage or disadvantage, depending on your point of view. But sceptics' arguments that you cannot feed a huge population with nut carbohydrate don't stand up. Plantations of modern nut varieties are much more productive than similar areas of arable crops. Wheat commonly produces between 2 and 10 tons per acre on good soils. On much poorer soils, chestnuts have an annual yield of 7 to 11 tons, pecans 9 to 11 tons, hazelnuts (most suitable for northern temperate regions) 9 to 12 tons, and walnuts 10 to 15 tons. In tropical areas, or in sites where the trees can be underplanted with vines or vegetables (or cereals, come to that), the overall yields can be far higher.

In the Mediterranean there are still thriving three-dimensional, 'forest-farming' systems which, against the odds, prosper in the supposedly unfriendly conditions. Most notable are those producing olives and cork oak.

Liquid gold

THE OLIVE CAN HARDLY BE DESCRIBED AS a staple, but it is one of the Mediterranean's most versatile and valuable plant crops. Eden's olive exhibit, with oil glistening at the bottom of glass artworks, is gushing in its hopefulness. Olive oil as the foundation of the fabled 'Mediterranean diet'. Olive oil as the sovereign remedy for the West's chronic illnesses, heart disease, cancer, arthritis. The captions quote Homer's description of it as 'liquid gold', and a thin, golden line weaves among the tiles on the pathway (though the ancients in fact liked their oil green, from olives picked just after the grape harvest).

Another olive-line marks the boundaries of the true Mediterranean zone. It encloses the area where the olive tree will grow happily, and closely follows the contour joining places with a mean February temperature of 7°C / 45° F. It passes through central Spain, southern France, the Italian lowlands, southern Greece and the Islands, the Holy Land, and back through northern Tunisia and Morocco. Olives will thrive anywhere the average winter temperature stays above 3°C / 37.4° F. The olive defines what we understand as the Mediterranean. It is what ecologists call an indicator species, though in this case it indicates patterns of culture as well as climate and vegetation.

Yet the variety of olive which bears edible fruits may not be native to much of the region. Its wild ancestor, the oleaster, *Olea europaea* ssp *oleaster* (rather oddly named as a sub-species of its cultivated descendant, and probably better named *O. sylvestris*), unquestionably is. This spiny, much-branched shrub with small, barely oily and inedible fruits grows in maquis, *garrigue* and hill country throughout the Mediterranean. Somewhere, at some moment (or more likely on many occasions), the oleaster must have thrown up a mutation, an altogether bigger plant, with larger leaves and a distinctly oily fruit. And on some other occasion, the art of grafting shoots of this sport on to wild olive stocks was learned. This is the only way the edible varieties can be propagated. If cultivated olives regenerate by seed, by crossing with oleasters, say, the fruit remains edible for a generation or two, but soon reverts to the wild form.

As with all anciently domesticated species, no one is certain where this first domestication might have occurred, though it looks as if the general movement of cultivated olives through the Mediterranean basin began in the east. Greece is ruled out by some commentators, because there is no mention of olives by early agricultural chroniclers, such as Hesiod, who was writing in the eighth century BCE. Syria is the favourite, because of the frequent mention of olives and oil in biblical sources that date from the same period. But these are all guesses based on written sources, which are notoriously unreliable on such matters. On archaeological evidence, Crete is as likely as anywhere to have been the focus of domestication, and at a very much earlier date than any of the suggestions from the Middle East. There is a noticeable increase in olive pollen from the Middle Neolithic onwards in Cretan excavations. Olive stones and possible prunings have been identified from the Early Minoan site in Myrtos. Cultivated olives were certainly about in Crete by 2500 BCE. In Spain, domestication may have happened independently, as wood seeming to come from cultivated rather than wild olives has been found in very early excavations.

Wherever the cultivated olive originated, it rapidly became one of the foundation stones of the Mediterranean economy. It could be successfully grown on the thin, infertile, limestone soils that dominate the region, and which made it difficult for many areas to become self-sufficient in wheat before the days of fertilizers. Indeed, the rise of Athens to greatness had a lot to do with its commerce in olive oil.

Olives were the currency of the Mediterranean. The oil was not just a fundamental part of the diet but a libation, an unguent, a therapy. Greek athletes used so much oil to anoint their bodies that there was a special curved knife called the *strigil* to scrape it off. The first Olympic flame was a burning olive branch, and olive wreaths crowned the victors.

The Bible refers to olive oil 140 times, and to the olive tree nearly one hundred. Olive oil was what Mary Magdalene rubbed on Christ's feet. The doors of Solomon's temple were built from olive wood. Muhammad likewise praised the oil, and likened the light from Allah's being to the radiance of a fuel from 'the Blessed Tree, neither of the East nor the West'. In all Mediterranean religions olives have symbolized peace. Yet because they are regarded, universally, as trees of the homeland, they have also become emblems of territorial aggression. Along Palestine's West Bank, both sides plant them to mark possession of the land. When gunfire, or even stones, fly in from behind the shelter of olives, Israeli bulldozers move in to flatten them.

The Greeks were responsible for spreading the cultivated olive west. They took it, most crucially, to Italy. They exported their skills in grafting and pruning to Sicily, and built Syracuse into a major olive-exporting port. In Tunisia olive groves were established by the fifth century BCE. It was via these north African groves that the cultivated olive arrived (maybe for the second time) in Spain. Moorish settlers introduced many more strains to Spain, and the Spanish words for olive and oil – *aceituna* and *aceite* – are both Arabic in origin. The

Spanish colonialists took trees to the New World in 1500, and Italian migrants carried them to the southern hemisphere in the nineteenth and twentieth centuries.

Today there are not far short of a thousand million olive trees on the planet, and more than one thousand named cultivars. China, surprisingly, has 20 million trees, which is four times as many as France, but 90 per cent of the world's olives grow around the Mediterranean. (And they 'can appear in unlikely places', writes Mort Rosenblum. 'In prewar Athens, fathers gave their pubescent sons a handful of drachma, a condom and the address of a good-hearted prostitute. My friend's father went to see Soula. Having no diaphragm, she relied on an alternative protection from pregnancy: a fat Kalamata olive.')

And as well as their role in the economies and cultures of the Mediterranean, they have helped shape the landscape, and views of that landscape. Aldous Huxley thought them the quintessential element of southern European painting: 'Under a polished sky the olives state their aesthetic case without the qualifications of mist, of shifting lights, of atmospheric perspective, which give to the English landscape their subtle and melancholy beauty ... [The olive] does not need to be transposed into another key, and it can be rendered completely in terms of pigments that are as old as the art of painting.' Van Gogh painted eighteen canvases of ancient olives, in deep greens and blues. Cézanne placed them in front of Mont Sainte-Victoire. Renoir painted them only rarely but was profoundly affected by them. 'Look at the light on the olives,' he wrote in his journal one afternoon. 'It sparkles like diamonds. It is pink, it is blue, and the sky that plays across them is enough to drive you mad.' At the beginning of the twentieth century, Renoir bought an old grove near his home in Cagnes, in Provence, that was due to be felled for charcoal. The trees are still there, wild, unpruned and thriving.

Eden's olives are a testament to the durability of the tree. Dominic Cole had told me the story of how they'd come to Cornwall. He'd found them in Sicily, growing in huge pots. They weren't as old as they looked, a hundred to a hundred and fifty years at the most, and had been pensioned off by the growers when their yields started to drop. But instead of being burned, the old trees were now being recycled to meet the growing demand for ornamental olives in gardens. These had survived being dug up, root-pruned, potted, carried by sea to England, having their roots entirely stripped of soil and washed to remove any possible pests, being repotted and repruned, and finally planted out again in the congenial climate of the Warm Temperate Biome.

The largest of the trees, close to the entrance to the biome, is a hulk of whorled wood. It is only six feet high and hard-pruned, but measures the same number of feet across its root buttress. The wood flows like lava into the stony soil. I run my eye slowly back up the trunk, following the wisp of Smilax that is winding round the bole. Every runner and low shoot that has been lopped off has left a mark, a dark, rounded pimple, whose scar tissue looks as if it had been turned on a tiny lathe. Halfway up the trunk is twisted, perhaps the result of a frost when the tree was young. Perhaps just some personal idiosyncrasy deep in its genes. The silver leaves are wide open. When the temperature is low or the wind is high, they fold up, holding their moisture in.

Olives die when weather conditions are very severe, but often it is only die-back, a shut-down of the vulnerable parts of the plant above ground level. A frost of −13° C / 8.6° F kills the tree right down to the ground. The base and root system survive, but the tree needs to have new grafts to produce edible fruit. The oldest olive trees (they can reach ages of at least two thousand years) contain records of great frosts in their grain-patterns. Olives in Provence were cut back to the ground by the great frost of 1709, but they survived and slowly recovered. There were more severe frosts in 1956 and 1963 (−29° C / −20.2° F), and again many of the trees might have recovered. But olive-growing was in decline and unsubsidized at that time, and many of the groves were abandoned, or grubbed out for alternative crops.

Olives have always been entangled in food hierarchies, their wildness or traditionalness counting for or against them according to fashion. A myth from southern Europe (not very ancient, one suspects) attempts to explain − and thus downgrade − the roughness of the wild olive tree. A group of nymphs, enjoying a dance, disturb a shepherd, who responds with catcalls and a crude parody of their dance. The gods don't think it a laughing matter and punish the shepherd by covering his head with bark and changing him into a wild oleaster. Today, it is precisely the apparent wildness and antiquity of olives that gives them their *mana*. Their 'liquid gold' is a distillation of Mediterranean sunshine and health, given credence by the 'wisdom' of the ancient, gnarled trees and their resilient roots.

In Australia, literal wildness is coming to be seen as a bonus in the olive population. One million new trees (mostly Italian cultivars not always well suited to the local climate) are being planted in Australia every year. The first trees were taken out to the continent by colonists in the early 1800s, and led to a short-lived olive oil boom in the 1890s. This didn't survive cheap imports and a cuisine not then very interested in exotic fruits, and many of the trees went feral, cross-breeding and producing semi-wild hybrids that seemed well adapted to Australian conditions. Indeed, they're now distinctive enough to be referred to by local connoisseurs not as feral but Feral, and are regarded as a unique genetic storehouse. In South Australia especially their intense flavour is preferred to that of the cultivated varieties.

Ferals grow along roadsides, and aren't always looked on sympathetically by tidy-minded local authorities. But a new wave of Mediterranean migrants has adopted them, and families gather the olives to press in reconditioned machines left over from the colonial olive days. Some of the Feral oils are available in specialist delicatessens, but like the distinctive local oils from all over the world, they have to battle continuously with cheap, blended oils from Italy.

The olives of Tunisia are proof that Mediterranean species can flourish in next to no soil. They are also an example of the way that the benefits of plants to humans may depend less on the plants themselves than on the social and political systems through which they're mediated. Those who've tasted Tunisian oils reckon them among the best in the world. The yield of the trees is prodigious: even on the edge of the Sahara Desert a giant tree of the *chemlai* variety can produce 800 kilos in a year. Its closest equivalent, the French *cailletier*, rarely tops 50 kilos across the Mediterranean region. But Tunisian oil is barely known. Like the oils from many other poor countries, it's shipped in bulk to Italy to be blended and relabelled as an anonymous, generic 'Italian' oil, and the Tunisians are switching, with government encouragement, to cheap, imported seed-oils for use at home.

North Africa may have been one of the other centres where oleasters were domesticated. The Phoenicians certainly brought olives here, as did the Romans, who began the tradition of importing oil from Jerba. During the 1950s, President Bourguiba introduced state olive groves, and by 1995, nine years after his death, there were 65 million trees in the country, something like 8 per cent of the world total. The trees seemed able to flourish almost anywhere, even in the Sahel, one of the most parched areas of the desert. They grow differently from olives in richer and damper soils. They put down tap-roots that often probe more than six metres underground to reach water. They become very tall and wide, to catch morning dew with their leaves. The Tunisian growers have gone along with the trees' natural inclinations. They plant trees at no more than six or seven per acre. They allow the branches to spread, and the picking is done from tall ladders. Tunisian olive-pickers use a unique tool, that may have existed for the past thousand years. They slip the ends of goats' horns over three fingers, and then harvest the olives by clawing at them. Neither the olive fruits' skins nor the fragile bark on the fruiting branches suffers. How long these traditions will survive is uncertain. Just as the character of many of the Mediterranean's local oils is being subsumed in blended products, so customs suited to local conditions and cultures are being swallowed up in European-wide plantation systems.

Dehesa, pairidaeza!

A<small>T THE HEART OF THE</small> M<small>EDITERRANEAN BIOME</small> is a grove of cork oaks, a representation of the great evergreen forests of south-west Spain and Portugal. It's the most beguiling display in the place. The oaks are underplanted with cistus and asphodel. Hunks of cork bark are scattered around for the visitors to squeeze. And rootling between the trees is a herd of pigs and piglets made from scraps of cork and driftwood by the artist Heather Jansch. They look warm and frisky, sniffing the breeze, burying their snouts in the ground.

Nearby, Mark, a regular Eden guide from Zimbabwe, is telling the story of cork. He's dressed in bush-hat and shorts, a convincingly warm and temperate man, and he's unravelling the story of how this anciently renewable resource is a linchpin for one of the last self-sufficient farming systems in Europe, the *dehesas* of Estremadura.

The cork oak is a classic Mediterranean tree. It's light-demanding and easily displaced if the savannas it favours progress to closed forest. It flourishes on poor, acid soils, especially on granite. It is just about evergreen, showing its 'autumn' colours in April and losing its old leaves as the new ones emerge. The extraordinary, insulating bark, which makes it such a uniquely valuable tree, is an adaptation to the Mediterranean's ever-present threat of fire. Cork oak is not very combustible in itself. An ordinary *garrigue* fire scarcely bothers it. A fiercer conflagration will burn the leaves and twigs, but simply provoke the trunk in its non-flammable casing to sprout new foliage. Such an adaptation suggests hundreds of thousands of years' exposure to natural fires. The cork is scorched by fire but not usually destroyed, and the cork-oakeries of Portugal are supposed to have evolved after the use of artificial fires to reduce the competing vegetation.

The bark itself is a honeycomb of tiny, watertight cells held together by a resinous substance called suberin. It's elastic, exceptionally light-weight, waterproof and fire-resistant. It is also one of the few tree barks that can completely renew itself when stripped away – seemingly indefinitely, though commercial cork growers generally regard a tree as past its best after 150 years (the equivalent of more than a dozen cycli-cal bark-strippings).

There are stories of the ancient Greeks using corks as stoppers for wine and olive jars, but these are almost certainly fictions, as the cork oak's eastern range does not stretch that far. But it was well known to the Romans, and used in beehives as well as bottles. The English word 'cork' originally meant a cork over-boot, a kind of galosh. The expan-sion in the wine trade in the sixteenth century meant a great increase in cork production, principally in Spain and Portugal. By the late eighteenth century France was importing most of its wine-bottle cork from Spain. In the nineteenth century a new use for cork was found as a constituent of linoleum, and this has remained the major use.

Yet the vast areas of open woodland in Iberia where the cork oak grows are anything but a simple cork-factory. I'd first seen these savannas myself one early spring in the late 1990s. Five of us had gone down to Estremadura, to watch the cranes that come here in winter to feed on the acorns from the cork and evergreen live-oaks (often mis-takenly described as holm oaks, *Quercus ilex*, though in fact the related Iberian species, *Q. rotundifolia*). We were based in Trujillo, the medieval hilltop town where Francisco Pizarro, conqueror of Peru, was born in 1475. He was to become a key figure in the spread of Western agricul-ture, taking European cattle to South America, and playing his own part in destroying indigenous farming methods. There is a statue to him in the Plaza Mayor, where we ate dark *jamón* from the Estremaduran pigs. Among the baroque arcades and palaces built for the returning conquistadors we could just make out the ghostly outlines of storks, perched like gargoyles on their nests on the walls and roofs. In summer I knew (and saw for myself the following year) that they would be joined by packs of swifts, lesser kestrels and red-rumped swallows. Trujillo seemed a town of birds as much as people, and a token of the easy and intimate relationship that exists between people and nature in this part of Spain.

But these intimations of spring dissolved the next morning. The Spanish plain's proverbial element arrived on cue. It rained for a few hours every succeeding day, a thin insidious drizzle that found its way through all our waterproofing, but covered the landscape with a film of fresh green. After an hour it scarcely mattered. We had driven into real *dehesa*, and it touched some deep memory in all of us. The dumpy cork and live-oaks formed an open evergreen forest, stippled with mossy boulders and cistus bushes. Cranes in groups of three or four, family parties, strolled about nibbling acorns. A Spanish imperial eagle perched democratically on an electricity pylon. Closer to us were flocks of azure-winged magpies, with sky-blue tails and pinkish bodies, garrulous birds, given to fanning swoops into the trees that showed off the best of their exotic plumage. In the middle distance storks padded across the grass, stabbing at frogs in the pools with their red dagger-beaks. We glimpsed a family of wild boar, and the antlers of a half-hidden herd of red deer.

It was easy to feel at home in this landscape, with its subtle mosaic of glades and groves, and we wandered about, sampling the sweet acorns ourselves – they're sold here as *bellotas* – and learning to swig wine from goatskin *botas*. I took to strolling about with my hands behind my back, feeling rather like the Duke of Edinburgh. Part of the familiarity of this place, I think, was the echo it struck of a park gone wild, or that archetypal forest-with-clearings landscape that occurs across the planet. These savannas occur wherever large herbivores meet forest. They are widespread in Australia, in the African bush, in South America, and throughout southern Europe. In Britain they are known as wood-pasture, and characterize areas like the New Forest and most large areas of commonland. In Old Persian, the word for a park or pleasure garden was *pairidaeza*, which led to our word paradise. The word comes from *daeza*, meaning 'wall', and *pairi*, 'around': already paradise was a fenced enclosure. The purpose of the garden was to recreate an earthly equivalent of the Heavenly Paradise, which was later described in the Qu'ran as a 'place of spreading shade' and 'fountains of gushing water' and abundant fruit trees and palms.

I thought of how deeply ingrained in our imaginations is this motif of an open forest with scattered trees and grazing animals and water. The biologist Edward O. Wilson regards it as the kind of habitat in which most of our species' evolution has happened. Forest clearings may have been the habitat where the crucial transitional stages between ape and human were made; and forest clearings are the dwelling space that all humans make when they first move into a previously unsettled wilderness. The memory of those primal human habitats, Wilson argues, may be very deep, genetically encoded. In his book *Biophilia* he argues that 'whenever people are given a free choice, they move to tree-studded land on prominences overlooking water.' The savanna may be a species memory, shared between all cultures.

Later we got the chance to see at close quarters how the economy of the *dehesa* works. There were nine of us in two vehicles now, as we'd been joined by three Spanish wildlife photographers, all dab hands with the *bota*. Our guide, Damian, was taking us to a remote farmhouse belonging to the Flores family. It was post-modern self-sufficiency, a big spread adorned with satellite dishes, an eco-ranch. The sons were down from university. We crammed into the tiny kitchen with the entire family. Grandma sat at a table in the corner with two daughters. Everything we ate had been grown or raised on the farm, and had been cooked in the same huge cauldron. A thin soup was dredged off the top, then a layer of chick-peas, and finally cryptic lumps of smoked meat and fat at the bottom. We passed round our hipflask of malt whisky after the meal, which was reciprocated with a tomato ketchup bottle full of an evocative lilac liqueur. It was sharp, aniseedy, blossom-scented, and we debated its identity with our hosts in a mixture of botanical Latin and street Spanish, too far from home to twig that it was just a local version of that Anglo-Saxon winter warmer, sloe gin.

After lunch we walked round the farm. It had begun to rain again, and the assembly of animals had taken on a bedraggled and prehistoric appearance. There were herds of multicoloured goats, Merino sheep, long-horned retinto cattle and multitudes of pigs and piglets,

some ginger like Tamworths, some dark brown *cerdo ibericos*, which are close to wild boars in ancestry. The *dehesa* system works through an intricate relationship between these grazing and rootling animals, the acorn-producing oaks, and the scrub and grassland between. The oaks, both live and cork, are lopped every eight or nine years to stimulate acorn production, and the cut wood is converted into charcoal for the European barbecue market. The acorns are eaten by the pigs in an impressive energy exchange. Each tree can produce up to 1,000 litres of nuts a year, which will grow 100 kilos of pork. The bark is stripped from the cork-oak trunks on an eight-to-ten-year cycle. The ground under the lopped trees is grazed by sheep and cattle and some of it ploughed up for arable crops every fourth year. And this symbiotic system, based on a habitat only marginally altered from its natural state (it is called 'forest farming' elsewhere), supports the spectacular wildlife that we had already seen.

Back in the temperate biome, Mark is finishing his story. Cork, as a commercial raw material, looked for a while as if it might be threatened by a shift in the drinking habits of the developed world. Drinkers these days are demanding predictable, trouble-free wine, to be consumed on the day it's bought. The subtleties of laying-down and ageing are part of yesterday's world. So the more reliable and thoroughly aseptic plastic stopper seemed to be edging out the natural cork. But the situation is already turning itself around. The rapidly expanding wine industries of the New World are returning to the cork (though there is a slow move towards metal screw-tops among top producers). And the cork industry has reacted to the threat with a flurry of research and development, leading to composite corks with better sealing properties. The *dehesas* did look under threat for different reasons during the 1960s and '70s, when the Spanish government, responding to conventional European agricultural thinking, tried to do away with the trees in favour of yet more cereal production. The experiment was futile, though possibly a third of the *dehesa* was destroyed during this period.

And the *dehesa*'s message may yet find its moment. Next to me, using the cork-oak grove and its sculptured pigs as a tableau, a father is explaining the system to his son, who has cerebral palsy. You cut off the cork, he explains, let it grow back, and then later strip it off again. The boy holds a tube of bark and moves it up and down and up again. That physical contact clicks in his mind, and I think he has grasped the great central principle of renewable resources: growth, harvest, regrowth. Will Eden be bold enough in return to take the next step, and demonstrate the harvesting of cork from its own trees?

The panacea

THAT DUSTY, RESINOUS SMELL of the south follows you through the whole biome, past the South African *fynbos* and the Californian chaparral, back towards the cistus and lavender again. Some eminent visitors have refused to believe the aroma is authentic, and have accused Eden of spraying ambient perfumes about the dome, as if it were some kind of department store. Today, there's a new but recognizable edge to the scent: the whiff of burnt wood. A new exhibit is being put together about the Mediterranean's old familiar, fire, and a pile of charred brush has been dumped on the bed which yesterday was occupied by San Lorenzo tomatoes.

By the exit the impact of this torrent of smells is acknowledged, and there is a whole exhibit devoted to perfume – though to tell the truth, it can't rival the tantalizing wafts of the *garrigue* behind you. There are pots of rather over-squeezed geranium and lavender, and a wall on which a long and serpentine tube, symbolically connecting the fragrances of plants with the human imagination, lassoes quotations about the potency of smell in lifting the spirits, evoking memories, carrying messages. Scent, after your long meander through some of its most inventive manifestations, begins to seem at the very heart of our relationship with plants.

In front of me a gang of rowdy French schoolchildren is whirling about by a quote from Coleridge: 'A dunghill at a distance smells like musk, and a dead dog like elderflowers.' I wonder what this canopied replica of part of their own country, Provence in a bottle, smells like to them? Against the wall there are three jars of some of the plant perfumes distilled in the south of France, arranged with inviting spouts so that visitors can sniff them. They're deliberately not labelled. The children place their noses extravagantly over the nozzles, pull faces, explode into giggles. Some put the nozzles *into* their noses. They gossip hilariously about boyfriends, teachers, farm animals. Smells, wrote Proust famously, 'bear unflinchingly, in the tiny and almost impalpable drop of their essence, the vast structure of recollection'.

I try putting my nose, a little more delicately, over the nozzles too. But they seem to have been sniffed dry. There is a hint of violet in one, and iris in another. But my own structure of recollection seems momentarily rather less than vast, and chiefly they remind me of some almost identical spouts in a perfume museum in Grasse, the centre of perfume production in Provence. I catch a hint of damp brass, the whiff of an evening perfume the morning after, a memory of disappointment at the drab display ... And suddenly the vast structure is unfolding after all, and much more pointedly than I could have anticipated: the unlocked memory was not of the vague smell of violets, but of one particular, exhibited, exhausted phial of violet in a museum.

Smell is our most direct, unmediated sense. It's a means of communication so ancient that we share it with the plant world, whose primary 'language' is made up of these streams of evocative chemicals. It is, of course, a language vastly older than words, and we can't describe scents except in terms of other scents. Even in the humble reaches of English wild flowers we're forced into extravagant and often incongruous similes. The flowers of Cornwall's rarity, the Plymouth Pear, reminded one sniffer of 'stinking scampi'. The wispy little umbellifer, stone parsley, smells, in Professor Tom Tutin's immortal phrase, like 'nutmeg mixed with petrol'. Indian balsam 'has a pervasive evening scent reminiscent of Jeyes fluid', and sweet-chestnut blossom, unequiv-ocally, of semen.

I thought about how, for me, the scents of spring flowers succeed each other as a narrative of memories. The little moschatel, or town-hall clock, has always been my first trigger. The first time I sniffed its modest five-faced flowers (the top face, my parent's generation used to say, was 'for the Spitfire pilots to read'), I was astonished to smell my first serious girlfriend, whose musk and almond fragrance had per-plexed me as a naive teenager. I imagined it was what a warm dormouse might smell like, and assumed it was some expensive per-fume. Now, probably only slightly wiser, I'm still transported back by moschatel to those first breathless clinches, and left puzzling about the strange evolutionary pathways that link the fragrance of a spring flower with a maturing young woman.

There was no moschatel in the Cornish lanes I walked through to Eden, but there were bluebells, whose scent is even more sultry. In small numbers I love the spicy, liliaceous overtone in their scent. But *en masse* I find them narcotic, a heavy haze that catches the throat. It's not just a physiological reaction. 'Wash wet like lakes', Gerard Manley Hopkins memorably described sheets of bluebells, and it's a feeling of drowning in their drowsy vapours that unsettles me – or, maybe, if I am honest, makes me feel too nostalgically smothered by one of the fundamental smells of my childhood. Either way I've come to feel about their scent much as I occasionally feel about their unmediated washes of colour: how good it would be if they rippled in a more dis-orderly way through the woods, shape-shifting, colour-rinsing, spicing the air as you walk (as they do often close to) with juggling mixtures of cinnamon and clove and balsam.

There are scents irrevocably attached to places, too. Lily of the valley means the limestone in the Yorkshire Dales one June, and a heavenly gust that hit me long before I saw the sheaves of clattering leaves. The smell of artemisia, *any* artemisia, takes me in a flash to the north Norfolk saltmarshes, where the bitter tang of sea wormwood blends with the wind and the call of whimbrels. Meadowsweet is the summer smell of my home in the fen country, a ubiquitous, astringent fragrance that rises up whenever you walk in a damp place, be it swamp or sedate riverbank. It has two scents: the fishy, sexy, may-like sweetness of the flower, and the almost medicinal cucumber and carbolic of the crushed leaves. Is this why one of its folk names is 'courtship and matrimony'? Eden's gorse, as I have said, has been a companion plant throughout my life, and its coconut scent recalls not just home but a host of other places and occasions. Watching peregrine falcons on a Welsh hillside, with gorse buds bursting in the heat so loudly that I couldn't tell them from the 'tchaks' of stonechats perched on the bushes. A natural gorse amphitheatre in Suffolk where I heard one of my first nightingales. Making an Elizabethan 'Grand Sallett' one New Year's Day with the first gorse buds.

Yet I'm remembering another kind of gorse experience, too. I'm lying on my back on a heath one June afternoon. The gorse and broom fronds are hanging like garlands against the sky. The smell is tropical – vanilla, melon, peach. But suddenly I'm hit by an extra burst of coconut. It's a pulse of scent, that seems to go not just up my nose but into my eyes and cheeks, and I wonder if a breeze has momentarily got up. But it's a day of dead calm. A few minutes later it happens again, and I recall noticing these rhythmic gusts with other kinds of fragrant flowers: with laburnums, lime trees, viburnums. I wonder if it's an olfactory illusion, a momentary shift of attention – something in me, not the plant. Violets do this, because of a substance called ionone, which briefly anaesthetizes our scent buds. The flower itself continues to smell but we lose the ability to register it. Then a couple of minutes later, the nose recovers and the smell reappears. But I don't think this is what is happening here. Might plants budget their precious scent molecules, emitting little puffs as come-ons? Does the proximity of other organisms influence their chemical goings-on?

Smell is probably not the oldest sensory system. The earliest cells must have first acquired a positional sense, an ability to orient them-selves in space and respond to gravity and warmth. But the identification of food, and then, crucially, the necessity of interaction with other organisms, entailed the interpretation of chemical signals and messages. This is the process known as chemotaxis (taxis means re-ordering or redirecting) which in essence is the prototype of the sense of smell. We can have problems imagining this transition, because smell to us is essentially a *conscious* experience. The notion of a smell that you cannot smell seems like an absurdity, a contradiction in terms. Yet an elaborate, and entirely unconscious, chemical signalling network was one of the legacies that simple organisms bequeathed to the animals that evolved from them. Long before humans began to register scent consciously, their behaviour was being guided by it. Smells helped with finding a mate and bonding with children and tribe, with locating food and avoiding danger, with interpreting news about the weather and the comings and goings of other creatures, and occa-sionally with providing pure, sensual pleasure. The smell receptors were the foundations of the limbic system, a primitive centre concerned with basic and powerful emotions, with the recording of sensations and the expression of desire, and it was round this that the apparatus of memory and reflection began to evolve. Our brains are outgrowths of our noses. No wonder that smells remain the great carriers and triggers of potent memories. They 'fix' emotions, and are processed in the same ancient area of our brains. They are mordants for experience.

Yet a gap has opened up between our ancestral sensitivity to smell and its role in our culture. As soon as humans started walking upright, and could see beyond their noses, sight began its long rise to ascendancy as the dominant sense. Today we like to imagine that smell is more or less superfluous as an aid to survival. We enjoy barbecues, Chanel No. 5, wine and roses, but beyond the rarities of gas leaks and putrefying food, we don't need our noses to stay alive.

However, the sense of smell refuses to wither, to become an evolutionary has-been. We still have three thousand genes encoded for smell, compared to just three for colour vision – a measure of just how dominant smell has been for most of life's history. They operate through a thousand discrete sense receptors, enable us to distinguish at least ten thousand different odours, and to mix and match these basic sensations into a huge lexicon of scent memories. We can tell the difference, with our eyes shut, between French lavender and spike lavender, vodka and gin, mist and smoke, spring and autumn. Does it make any sense to ask why we should be able to do this? Can there possibly be an evolutionary reason why we have a receptor for, say, the smell of nutmeg? Or is it just a happy molecular coincidence, analogous to the fact that cats seem to have a receptor plumb in their pleasure centres for catmint, for which it's doubtful whether they've ever had any kind of functional need?

For all its overlay of cultural associations, smell is the most direct of our senses. The odour molecules fly straight to the receptor cells at the back of the nose. And there is no way, except in the very short term, we can close scent down, in the way we can sight. To stop smelling we would have to stop breathing. Smells continue to operate below the levels of our consciousness. The fact that women living together in a dormitory begin to synchronize their menstrual cycles as a result of airborne hormones is now part of popular folklore. Occult smells may influence our choice of partners and friends, and the smoothness of social gatherings. In Patrick Süskind's nose-gripping and blackly comic fantasy, *Perfume*, set in eighteenth-century France, the hero is a man born without any personal odour whatsoever. But he has an exquisite sense of smell, which leads him to become an almost supernatural concocter of perfumes. Distillation is what fascinates him most. The production of that 'scented soul, that ethereal oil, was in fact the best thing about matter, the only reason for his interest in it. The rest of the stupid stuff – the blossoms, leaves, rind, fruit, colour, beauty, vitality and all those other useless qualities – were of no concern to him. They were mere husk and ballast, to be disposed of.' But his lack of a human scent disturbs other people, so he devises fragrances to make him socially acceptable, 'an odour of inconspicuousness', a hint of new-born infant, and eventually a smell so irresistible that it brings about his demise in the most cataclysmic of ways. (He is eaten by a mob.)

This borderland between the natural and cultural impact of smell is where the perfume industry operates, and has seen the emergence of everything from room fresheners which supposedly smell of spring woodland to synthetic human pheromones to spice up the wearers' sex lives. There is nothing new in this. One of the first and most graphic evocations of the power of scent is the Bible's Song of Solomon, the intensely erotic duet between Solomon and his lover in which smells and blossoms become both seductions towards, and symbols of, their physical love. It's set in a garden which echoes the original Eden, a *hortus conclusus* full of fragrant plants. 'A bundle of myrrh is my well-beloved unto me', Solomon's lover sings, 'he shall lie all night betwixt my breasts ... As the apple tree among the trees of the wood, so is my beloved among the sons. I sat down under his shadow with great delight, and his fruit was sweet to my taste'. Solomon replies: 'Thy lips, O my spouse, drop as the honeycomb: honey and milk are under thy tongue; and the smell of thy garments is like the smell of Lebanon. A garden inclosed is my sister, my spouse; a spring shut up, a fountain sealed. Thy plants are an orchard of pomegranates, with pleasant fruits; camphire [the henna-plant], with spikenard, spikenard and saffron; calamus and cinnamon, with all trees of frank-incense; myrrh and aloes, with all the chief spices'. She responds: 'Awake, O north wind; and come thou south; blow upon my garden, that the spices thereof may flow out. Let my beloved come into his garden, and eat his pleasant fruits ... My beloved put his hand by the hole of the door, and my bowels were moved for him. I rose up to open to my beloved; and my hands dropped with myrrh, and my fingers with sweet smelling myrrh, upon the handles of the lock.'

It's impossible for us, reacting to smells, to disentangle which part of our response is physiological, which symbolic, which a result of cultural associations. When *National Geographic* magazine did a global survey of human smell sensitivity using scratch-and-sniff forms, it found that almost all its one and a half million respondents identified and enjoyed the scent of roses, regardless of gender, race or cultural background. Has the rose, now a component of every garden and open space, become a universally familiar scent? Or has it some chemical characteristics that speak directly to the pleasure centres of the brain?

If scents are potent agents for fixing and unlocking human memories, they also make up the collective memory of the plant world, the language through which plants communicate with each other and with their surroundings. The chemical signalling which goes on between plants and the soil, other organisms and other plants is ceaseless. From the moment a seed enters the ground it begins releasing compounds which nurse and protect it, chemicals which suppress the growth of other seedlings and micro-organisms in the vicinity. They aren't long-lived; just sufficient to give the developing plant a start. Seeds also have receptors for 'smelling' the chemistry in the soil, so that they can start germination in the right conditions. But once a plant's root system is established, other kinds of bioactive substances are released into its surroundings. These help promote the growth of symbiotic fungi, and of nitrogen-fixing bacteria – which are, as it were, smelling these chemicals in their turn.

Flowering plants also use volatile chemicals to 'call in' pollinators. These can be active in astonishingly low concentrations. Fruit flies will respond to as little as one hundredth millionth of a gram of a pheromone produced by *Cassia* plants. Honey bees are able to read and interpret the chemical cues diffused into the atmosphere over a range of maybe 40 square kilometres, to remember the location of nectar-rich plants and subsequently convey the information to other bees from their colony.

Conversely, when visited by damaging insects, some trees release chemical messengers into the air which, when they reach other trees, turn on the production of tannin and other astringent and indigestible substances which deter the insects from feeding. The messenger triggers the trees' memory. In response to overfeeding by aphids some plants release a volatile aromatic from their leaves, which mimics an aphid alarm pheromone warning of approaching predators, telling them to flee the plant. Lima beans affected by spider mite give off volatile terpenoids which attract another species of predatory mite that feeds on the original spider mites. But not a random predator; the bean analyses the spider mites' saliva and releases scented chemicals which 'calls' only the species that feeds on them.

And it looks as if, in established ecosystems, there is a constantly interacting rain of volatile chemical compounds falling from the leaves on to the soil, helping to balance the composition of the vegetation, preventing overcrowding, encouraging the development of moisture-conserving ground-cover plants and symbiotic fungi. Lewis Thomas had a vision of the entire planet self-regulated by its smells. 'In this immense organism, chemical signals might serve the function of global hormones, keeping balance and symmetry in the operation of various interrelated working parts, informing tissues in the vegetation of the Alps about the state of eels in the Sargasso Sea, by long, interminable relays of interconnected messages between all kinds of other creatures.'

It would be good if we could be part of that too.

In many pre-scientific cultures it has always been assumed that we *are* part of it. Plants, it's believed, signal to humans through their complex structures and ways of life, and inform us about their properties. Their smells, shapes, times of flowering, their companions and the places in which they grow, all give clues as to how they might affect human beings, especially in illness. This is the widespread doctrine of sympathetic magic, crudely described as the idea of 'like cures, or in some way influences, like', but in reality a complex and subtle system for predicting the way that plants might react with human physiology. In modern scientific terms it is of course nonsense, but it was based on a view of the connectivity of living things.

In late sixteenth-century Europe, a rather precious version of sympathetic magic became fashionable under the name of the Doctrine of Signatures. In one respect this was another long shadow of the Eden myth. People had always argued about whether there had been diseases before the Fall, whether the Garden contained medicinal plants, whether illness was one of God's punishments on humans. But it was widely believed that God would not have abandoned us without at least the possibility of remedies. The problem was a practical one, of divining which plant was intended for which disease, to 'Restore to us, in part, what Adam knew before'. The favoured solution was based on the idea that God had 'signed' all plants with indications of their curative properties – as if he too were an adherent of sympathetic magic. The most evangelical proponent was William Coles. His major book was called *Adam in Eden, or Nature's Paradise* (1657), and in the introduction he gives an elegant outline of signaturism:

> Though Sin and Sathan have plunged mankinde into an Ocean
> of Infirmities Yet the mercy of God which is over all his Workes
> Maketh Grasse to grow upon the Mountaines and Herbs for the
> use of Men and hath not onely stamped upon them (as upon
> every man) a distinct forme, but also given them particular
> signatures, whereby a Man may read even in legible Characters
> the Use of them. Heart Trefoyle is so called not onely because the
> Leafe is Triangular like the Heart of a Man, but also because each
> leafe contains the perfect Icon of an Heart and that in its proper
> colour viz a flesh colour. Hounds tongue hath a forme not much
> different from its name which will tye the Tongues of Hounds so

that they shall not barke at you: if it be laid under the bottomes of ones feet. Wallnuts bear the whole Signature of the Head, the outwardmost green barke answerable to the thick skin whereunto the head is covered, and salt made of it is singularly good for wounds in that part, as the Kernell is good for the braines, which it resembles being environed with a shell which imitates the Scull, and then it is wrapped up againe in a silken covering somewhat representing the *Pia Mater*.

In the seventeenth century, decoding signatures rapidly degenerated into something resembling a parlour game, and enormous ingenuity – not to say self-delusion – was employed spotting panaceas in the shapes of pods and the suggestive curls of petals. The white-spotted, baggy leaves of lungwort were given for tuberculosis. Maidenhair fern, 'a long Moss that hangs upon Trees', was prescribed for baldness, and the knobby roots of celandine for piles. Nor was it just the external appearances of plants that were dredged for meaning. So were their habits. Stonecrops, whose roots could penetrate walls, were thought to be capable of effecting the same trick on kidney stones. The seed-heads of quaking grass, with their habit of trembling in the wind, were given for the 'shaking palsy' of Parkinson's disease.

Things soon got out of hand, and different interpreters saw different signatures in the same plant, one looking, say, at the shape of the strawberry fruit (wart-like), another at its colour (blood-like). In the case of poisonous plants the recommended signature, bewilderingly, often pointed out the plants to avoid, not to use. When the doctrine invaded the normally more earthbound world of gardening, it ought to have been time to call a halt. But Thomas Hill, in the first popular

gardening book, *The Gardener's Labyrinth*, passed on the farcical suggestion that a few lentils should be sown in vegetable plots as a guard
against the wind.

The Doctrine of Signatures' most significant difference from traditional sympathetic magic was its intense human-centredness. The
signatures weren't accidental clues which could be cracked by field-
work or – more drastically – by trance-induced dreams. They had been
designed, specifically for the benefit of the human species. The shape of
the walnut's shell was intended not so much to perfectly contain the
walnut, as to remind us of our brains, and their vastly superior powers.

The doctrine continued to exert a powerful influence on plant medicine, and, extraordinarily, still does today, especially on medical
herbalism and homeopathy. And in the seventeenth and eighteenth
centuries it went with the European settlers to the New World, along
with other varieties of folk and popular medicine. Vestiges of it still survive in the relatively isolated regions of the Appalachian mountains (as
do echoes of seventeenth-century diction and folk music). But the
Appalachians isn't some cultural reserve frozen in time. Its folk medicine today is a lively mixture of traditional beliefs, signaturism,
sympathetic magic, tried-and-tested 'family remedies', cures borrowed
from native Americans, and ideas picked up from popular medical
encyclopaedias. (These, including Jethro Kloss's *Back to Eden: Healing
Herbs, Home Remedies, Diet, and Health*, continue the long practice of cash-
ing in on the Eden myth.) Sometimes the origins are clear. Peach tree
root-bark is a remedy for diarrhoea, but must be gathered by scraping

the bark sympathetically *upwards*, against the direction of the runs. But is running a pair of scissors down the back to cure a nose-bleed also done because of the principles of sympathetic magic ('cutting' the blood) or just because something cold might stop the flow?

There isn't any evidence that the first settlers made use of the local herb ginseng, *Panax quinquefolium* – the one medicinal plant to have acquired the official tag of a panacea. (The genus was named *Panax* in 1753 by Linnaeus, who clearly knew both the myth of the panacea and ginseng's herbal reputation.) But the Appalachian Cherokees (whose theory of the origin of diseases, incidentally, was that the animals had created them in retribution for the lack of respect they'd been shown by humans) certainly did. Its roots were a cure-all, made into a tea and given for cholera, convulsions, colic, earache, fever, gonorrhoea, laziness, rheumatism and vertigo, but perhaps most commonly for respiratory problems and weakness.

American ginseng is a member of the plant group that includes ivy. It grows in shaded woodland right down the eastern coastal states from Quebec to Florida. It's a modest plant, ten to twenty inches high, with five-lobed leaves and small yellow flowers. Its most significant feature, from a herbalist's point of view, is its root, which is thick and forked, sometimes into the same fanciful resemblance to the human form that was once seen in the mandrake. This is a feature it shares with its better-known cousin *Panax ginseng*, Chinese ginseng, which has been used in traditional Asian medicine for maybe fifteen hundred years. The list of ailments for which this has been recommended is as all-embracing as that of its American cousin, and again features respiratory illnesses and debility. It was also believed to increase sexual energy, and it may have been this that caused the wild Chinese plants –

believed to be the most potent – to become rare during the eighteenth century. The price rose to ten times that of the same weight of gold, and encouraged the first experiments in cultivation.

In the early eighteenth century, Jesuit priests 'discovered' American ginseng, and over the next hundred years built up a profitable business gathering the herb from the wild and exporting it to China. During the depression of 1857–8, when many Appalachian farmers and smallholders went bankrupt, ginseng became an invaluable wild cash-crop. Whole communities turned to hunting the herb, and there are stories of local native American tribes getting in on the act, too. It helped tide over the local economy, though in many places ginseng was driven close to extinction, and some not very successful attempts at cultivation were started.

Today, ginseng populations have recovered a little, and digging for roots – known locally as ''sanging' – remains a vital part of Appalachian community life. Collectors receive an average of $450 a pound, and during the mid-'90s almost 100,000lb were gathered annually in the states of Kentucky, West Virginia and Tennessee alone. But this is a minute amount compared to the quantities of Asian ginseng that are cultivated in Korea to supply the vast world market that has opened up since ginseng became fashionable as a complementary tonic – a market that now sees extracts of the plant appearing indiscriminately in soft drinks, teas and beauty products in a way that seems in line with its old reputation as a panacea.

Does the panacea work? The answer would appear to depend on where you live. Different cultures find ginseng's effects very much in accord with their expectations. In Russia, Chinese ginseng (or more probably the unrelated Siberian 'ginseng') was given to cosmonauts to increase their stamina. And perhaps under Russian rather than traditional Chinese influence, the Viet Cong used ginseng to treat gunshot wounds during the Vietnam war. Orthodox Western medicine has tested ginseng and found it has no measurable effects beyond an occasional rise in blood pressure – except that in a typically unpleasant laboratory test, rats swimming in an escape-proof tank took longer to drown when they were dosed with ginseng than when they weren't.

Practitioners of complementary medicine have invented the term 'adaptogen' to describe ginseng's elusive effects on human physiology. It helps the body adapt to stress of all kinds, they argue, normalizing energy levels, immune responses, appetite and mood – which is why its effects are so hard to quantify. It sounds like an easy way of avoiding the discipline of serious testing, but it might be true. Such chemicals are well known in the plant world. One is salicylic acid – the precursor of aspirin. It was discovered in willow, but is widespread in plants. It appears to act botanically as an anti-inflammatory, anti-stress and growth-promoting substance. Its action on humans isn't identical (we use aspirin most often for pain, for instance, which is a phenomenon plants don't seem to have the mechanisms to experience), but there is enough overlap to remind us that our evolutionary roots lie in the same cells as those of plants. Because of this shared ancestry there is always a chance of finding compounds in plants which will be therapeutic in humans; we are simply dipping into the common legacy of biological self-medication. And the chances are that it will be a quicker route than the more random path of testing synthetics. (That Prozac was originally known as compound 82816 in Eli Lilley's laboratories gives a slight taste of how long this path can be.)

But it isn't infallible. We are not plants. Our bodies have physiologies which are uniquely mammalian and human. Compounds that are therapeutic in plants can be lethal in our systems, and some of the most effective drugs (e.g., atropine, from deadly nightshade) may be plants' waste products, or a kind of accidental chemical ornamentation, like the 'edge' of a scent. And it's worth remembering that plants invented 'magic bullets' for their own purposes – a scent, for instance, to deter a single species of predatory aphids – before we assume, with our usual human-centredness, that panaceas for our convenience will always lie waiting in their unexplored depths.

In Lucknow in Uttar Pradesh, there is a small but thriving industry that makes a perfume called *matti ka attar*, which roughly translates as 'earth scent'. It sounds an unappealing cosmetic, were it not for the fact that *matti ka attar* is that most evocative of aromas, the smell of new rain on parched earth, that indefinable hint of rising sap and autumn mellowness. Its manufacture is a simple and ancient process. The perfume-makers leave discs of clay exposed to the hot sun throughout the hot months of May and June. Well before the rains come, the discs are broken up, distilled in steam and their volatile aromas collected by absorbing them in sandalwood oil.

In the early 1970s, this traditional craft came under the scrutiny of workers from the Division of Mineral Chemistry of the Commonwealth Scientific Institute of Research Organizations in Melbourne. They set themselves the challenge of working out the origin and chemical composition of this mysterious emanation, which they called, rather poetically, 'petrichor': the essence of stone.

Their quest wasn't entirely original. Back in 1891, the French chemist Marcellin Berthelot concluded that the 'argillaceous odour' came from the residues of dead plants in the soil, revivified by water. A perfume scientist in the 1930s thought that it originated with fungus spores, especially from the Actinomycetes group, from which perfumers extract essences to mix with their scents. But the Australian team, I. J. Bear and G. R. Thomas, soon discovered these early ideas were wrong. In their laboratory they perfected a version of the techniques of the Lucknow alchemists. They took basalt clinker from the core of an extinct volcano in Victoria, granite gravel from Melbourne, quarry rubble and mine-spoil. They roasted all these varieties of stone to rid them of dead vegetation and fungus spores. Then they laid them out in the Australian sunshine for a season, to soak up whatever came their way. Finally each one was distilled in steam, to produce a few drops of oily fluid – which, regardless of the type of stone, unfailingly smelled of rain falling on dry earth.

The precise composition of petrichor eluded Bear and Thomas, though they found plenty of volatile hydrocarbons in their distillate. They followed up other leads, including a series of reports from the US National Academy of Sciences which showed that the blue haze that shimmers over vegetated landscapes on hot days is also made up of volatile essences from leaves and grass. The authors of this study had gone one stage further, and calculated that every year the vegetation of the planet is distilling 438 million tons of oil into the air. This is the source of petrichor, and the reason for its evocative appeal. It is the sum of the essences of Provence lavender and cedars of Lebanon, of lemon zest and orange blossom, of waxes from leaf cuticles and beads of nectar from bee-balm flowers, of volatile scents from eucalyptus trees in the deserts of Australia and balm of Gilead poplars in Europe, of waterproofing leaf-oils and fugitive plant pheromones, of insect repellents and insect attractants, of frankincense and myrrh and cinnamon and aloes and 'all the chief spices'. Released back into the air by the rain, petrichor has no 'purpose'. It's a mixture of random chemical waste and the remains of that deluge of volatile chemicals which plants pour on the earth to regulate soil ecology and seedling growth. The pleasure and torrent of associations it provokes in us are purely fortuitous.

Then the Australian scientists made an audacious suggestion. Against the conventional assumption that the oil deposits of the earth are, like coal, the products of the long compression of immense quantities of buried plants, they proposed another model. That this vast oily distillate of the world's vegetation, subsequently absorbed by porous rocks at the rate of billions of tons every century, might over the long course of geological time be the origins of the petroleum deposits in the hot regions of the world. And if so, that it was a process which is still going on: the waftings from Solomon's garden were on their way to the oil wells of the war-ravaged Middle East.

It is hard to know how to handle information like this. Would our attitude to oil, as the noxious substance at the heart of the planet's ills, change if it was proved to be a renewable resource, like wood? Might we use this process as an argument for conserving the vegetation of the dry parts of the planet? Maybe it would be wisest to view it simply as an example of the complexity of the plant world, and confirmation that this world does not work with us in mind. As Coleridge remarked: 'A dunghill at a distance smells like musk.'

THE
SECOND EDEN

Out of the strong

AS YOU WALK THROUGH THE SWING DOORS into the Humid Tropics Biome on a summer's day, you're hit by a blast of air. But it's not coming *at* you, as you might expect, gasped out of this hot cauldron. Instead you're being sucked in, borne on the cooling currents that flow in to replace the rising miasmas of the tropics.

This must have been how the first European explorers felt as they carried the fresh air of enlightenment and religion to the new worlds. They had complicated missions. They were, at root, after material riches: spices, gold, unimagined food crops. They had a direct religious project, too, to carry on the global crusade against Islam, which they believed could be accomplished via the 'back door' – by crossing the Atlantic. But many also were convinced that they would find the literal Garden of Eden, somewhere far away in the east.

A few of those who took the story of Eden literally in the Middle Ages believed that Eden had been irrevocably swept away by the Flood, and of these, some, in thrall to the symmetry of Christian religion, thought that the crucifixion had taken place on the exact site of the Fall. This, as John Prest has described, found expression in the legend of Seth, 'who was credited with having made his way back to the Garden, where he begged seeds from the tree of life off the angel on sentry duty at the gate, which he planted in the mouth of his father Adam, when he died. From these seeds, the cedars of Lebanon were, with a pleasing sense of historical continuity, thought to have been derived, and from the wood of one of the cedars of Lebanon felled for the construction of Solomon's Temple, after many miraculous adventures, the cross itself was supposed to have been hewn.' (Episodes of

Seth's labyrinthine story have had some strange modern echoes. In 1913 a fig tree sprouted from a grave in Watford Cemetery in Hertfordshire. It's known locally as the Atheist's Fig, and the story is that it grew from the tomb of a non-believer, who asked for a fig to be placed in his mouth or hand, and declared that if there was life beyond the grave it would spring to life.)

But by the late medieval period, it was more or less taken for granted that the Garden had survived the Flood and existed somewhere on earth (or, just conceivably, on the moon). The dawning realization that, in more southerly regions, the contrast between the seasons weakened, strengthened the belief that the place of perpetual spring could be found. When Columbus landed in South America on his third voyage in 1492, he wrote to the Spanish sovereigns who employed him, convinced that he had discovered the Earthly Paradise. This was no metaphorical claim. Columbus laid out a string of theological arguments about why this must be the true site of Eden. He thought he was at the extremity of Asia, where the sun rose on the day of creation. There was an abundance of fruit, the natives wore gold, there were four rivers . . . Yet in the end his case collapsed. There were too many unaccountable and discordant features in the new land: plants and animals for which there were no biblical names, tribes that practised cannibalism and incest. The discovery of the passion flower seemed a sure sign of God's handiwork, but where were the olives and the wheat? As for the anaconda, even the most penitent explorers felt that it was overdone as a model of the Serpent. The idea of a physically surviving Eden evaporated, and hereafter it would have to be rebuilt, according to the colonialists' agendas.

Meanwhile, the Portuguese had invented the ocean-going carrack, and were edging south down the west coast of Africa, moving ever closer to the climate zone of the perpetual spring. In 1419 they reached Madeira, an uninhabited island. They established a settlement, and the first boy and girl to be born there were christened Adam and Eve. Madeira was covered with extensive – and highly flammable – evergreen forest, which the new colonists proceeded to burn down. The large number of pigs and cattle they imported ensured that the trees would be slow to regenerate. Meanwhile they were establishing beehives and wheatfields, and grapevines brought all the way from Crete.

But this was subsistence farming, and not the reason the settlers had made the perilous journey from Portugal. They needed a crop to make their fortunes, and they decided on sugar. In the humid climate sugar-cane grew like a weed, and as early as 1452 the Portuguese crown authorized the first sugar mill on the island. In 1456 the first Madeiran sugar was exported to England, and by the first decades of the next century output was over a million kilograms a year. Ships were carrying sugar to Venice, Flanders, even Constantinople. What had happened to sugar, scarcely known in Europe two centuries before, to make it such a yearned-for product?

Sugary substances have been sought out round the world since at least Palaeolithic times. Their sweetness seems to hit a biological receptor in most animals – perhaps because it is usually an indicator of a high-energy food. Throughout most of the world, honey was the most ancient, widespread and popular source of sugar. It had the great advantage of being not just a fuel but a highly agreeable food in its own right, especially bitten into from a full honeycomb.

But there were many other sugar-providing foods. The Romans boiled down grape juice to syrups of varying concentrations. *Defrutum* was juice reduced by one half. *Passum* was the thickest, a sticky raisin wine. The Arabs did the same with dates and figs. (Date syrup is still widely available commercially.) The Chinese worked out how to use the sap of sagwire, a member of the palm family, and native Americans were blessed with the sap of maple trees, which they tapped and concentrated into a syrup which is still the tastiest alternative to honey. Britain had to make do with boiled parsnips – though medlars and wild-service-tree berries both produce a sweetish pulp when allowed to rot, or 'blet' as it's called. Up until well into the twentieth century, in the Kentish Weald children who wanted a winter sweetmeat picked a service-berry off a string by the hearth.

But somewhere in central Asia, the juice of a large member of the grass family was the favoured sugar source. No one has ever discovered the natural home of sugar-cane or found it growing in a convincingly wild form, though there is circumstantial evidence that it originated somewhere in India. The first Europeans to encounter it were probably Alexander's troops. Pliny wrote, a little vaguely, that 'it is a kind of honey that collects in reeds, white like gum and brittle to the teeth, the largest pieces are the size of a filbert. It is used only in medicine.' There is some evidence that it had been in cultivation there for a long time (the root of most 'sugar' words in Europe is the Sanskrit *sakkara*, for instance), but little is known about how it was grown or used.

It was the Arabs, again, who were responsible for spreading the sugar-cane westward, during their expansion in the seventh century AD. They established early and small-scale plantations from Egypt and Cyprus to north Africa and Spain – and almost certainly employed slave labour in them. The first sugar to be tasted in Britain was brought back from Syria by returning Crusaders in 1099. It was, at this stage, prohibitively expensive and regarded largely as a medicine (and, as with many profitable crops, its virtues as a panacea were talked up by its producers). But for those who'd tasted it, it began to seem like an irresistible luxury. An account-book from Henry III's household in 1226 refers to the procurement of 3lb from the Mayor of Winchester, and by 1288 the Royal Household was consuming 6,000lb a year. Thirty years later, its price in London was 2 shillings a pound, equivalent to perhaps £10 today.

But by the time the Madeiran plantations had swung into full pro-duction the demand for sugar had escalated and its price was dropping fast. The Portuguese began using slaves almost from the outset, in the herculean task of clearing the stumps from the burnt forest, shoring up cultivation terraces on the steep slopes, and creating 700km of irriga-tion channels and tunnels to bring water down from the mountains (Madeira is a mere 60km long). The work in the mills was also physi-cally gruelling. The canes were initially fed into a large set of rollers, to squeeze out the juice. It was a dangerous process, and there were many cases of workers getting pulled into the rollers. The brown syrup was then piped down to a cistern in the boiling house, where it was ladled – by hand – into kettles above the furnace. This process was repeated several times, using coppers of decreasing size but keeping the liquor at progressively higher temperatures, until it was concentrated enough to crystallize on cooling. (1 gallon of sugar-cane juice yielded 1lb muscovado sugar.) Daytime temperatures in the boiling house often reached 60° C / 160° F, with no ventilation, which is close to the tolerance level of the human body.

By the late fifteenth century, there were more than two thousand slaves on the island. There were not many blacks at this stage (the trade from the west African coast had only just begun) and most were either Berbers from north Africa or Guanches from the Canary Isles, who had entered the currency of European slavery long before Madeira was settled. The Guanche were probably the first people to be driven into extinction by Western imperialism.

The Madeiran experiment was so successful (in the colonizers' terms, anyway) that it went on to form the template for the plantation system worldwide, and for the future development of large-scale, and eventually industrial, agriculture. All the components were in place early on: the seizing and rationalization of land, the destruction of natural habitats and indigenous cultures, the establishment of crop monocultures, the employment of a cowed labour force. And all this done not in the interests of the local population but of a distant market, clamouring for what was not even an authentic food crop but a luxury.

Sugar plantations spread rapidly once Europeans had a foothold in the New World. Columbus carried sugar-canes on his second voyage, and planted them on the island of Hispaniola. In 1515, the canes' descendants produced the first cargo of New World sugar to be shipped back to Seville. As the Spanish and Portuguese influence began to wane in the Caribbean, so the British moved in. The first settlement was on St Kitts, in 1624. Barbados, which was to become the centre of British sugar production, was colonized in 1627, and within fifty years it had become one of the world's most densely populated agricultural areas, with some 40,000 people (two-thirds of them white) living on 430 square kilometres. By 1700 there were 900 sugar plantations on the island, and they occupied 80 per cent of the available arable land. Twenty-five years later there was no uncultivated land left. The sugar barons had created the world's first and most comprehensive monoculture on a country-wide scale, and with it came all the problems we now associate with intensive agriculture, including disease, soil deterioration and erosion. In addition, sugar-growing made it difficult subsequently to grow other crops in the same place. All this was achieved almost entirely by the use of slave labour. By the end of the seventeenth century, Britain had already forcibly transported over a quarter of a million Africans to the Caribbean islands. Between 1662 and 1807 – the year that Parliament outlawed the transportation of slaves – Britain had shipped out 3.4 million.

In many ways, the plantations in the New World were a test-bed for the agricultural revolution that was to sweep across Britain, and later most of Europe, in the eighteenth century. They proved the economic efficiency – though hardly the desirability – of a way of growing crops that was in dramatic contrast to the low-key, comparatively self-sufficient systems that had been the norm in the West for several thousand years.

The beginning of the nineteenth century (when British consumption had risen to 20lb a head) saw a significant benchmark in the rise of cane sugar. This was the moment when the price of sugar first fell below that of honey. In 1250 honey was fifty times cheaper than sugar. In 1600 it was still four times cheaper than a crop which had to be hauled all the way across the Atlantic. What happened to make this seemingly benign and freely given crop fall out of favour?

Honey – along with milk – is the only substance eaten by humans which is made specifically to be a food, albeit for the benefit of another species. It was, as we've seen, gathered by Palaeolithic people. Modern hunter-foragers still regard it as the most delectable and precious of foods, and eat it on the spot, honeycomb, dead bees and all. It was the food of the gods, *amrita* in India, ambrosia in Greece – and not just because of the instant explosion of fragrance and sweetness that sucking on it provides. It was also a magical food, found in the habitations of bees but, surely, brought by them from some higher place. Pliny, normally a level-headed naturalist, wondered if honey was 'a saliva emanating from the stars', or 'a juice exuding from the air while purifying it', or perhaps 'the sweat of the heavens', but he was certain that it descended from the skies: 'This substance is engendered from the air, mostly at the rising of the constellations'. Alas, because it fell from 'so vast a height', it was 'tainted by the exhalations of the earth' and by being 'accumulated in the stomachs of bees, and then cast up again'. He thought that bees were the carriers of the food, not the makers of it.

And in an oblique way he was right, though not about honey's celestial origins. Although it is processed by animals, honey is arguably the purest and most concentrated of all plant foods. It is the essence of flowers. The transformation wrought by the bees was not understood until the nineteenth century, but it proved to be comprehensible, and to owe more to diligence and delicacy than magic.

Nectar is thought to be the gift of the flower to the bee, the tiny bribe by which insects are rewarded if they perform the rites of pollen-transfer. Yet nectar occurs in species that don't rely on insects for fertilization, and it may have other purposes of which we're not yet aware. It lies, almost invisibly, in tiny pockets at the base of the petals called nectaries. The female worker honey bee heads for these, and sips their juice into a transparent bag that lies at the front of her abdomen. When this is full she heads back to the hive. In the hive, she and the other foragers pass the nectar on to the house bees. Pliny was right about 'saliva' at least: each bee regurgitates the nectar into drops on the underside of her proboscis, and the exposure to the air helps evaporate some of its water content. The nectar is passed on from bee to bee until it is concentrated to 40 per cent of its original moisture level. Then it's spread across the wax comb, where the warm air in the hive (several degrees higher than the ambient temperature) evaporates more moisture. All the while, the happenstance draught of air from passing foragers' wings, and deliberate fanning by other workers, reduces the liquid down to pure honey. When this is done, the bees cap the cell with wax, ready for when they need it – or for when it is scrumped by humans.

But this bare description of the process hardly does justice to the sheer variety of honey. The character of the individual flower is retained in the liquor. Hattie Ellis writes:

> The tree of heaven's honey tastes faintly of muscat grapes. Bees fly between orange blossom and the splayed white flowers of the coffee plant, fusing their flavours as they go ... A slightly salty, snow-white honey comes from the pohutukawa, the Christmas tree of the blazing, antipodean midsummer, which flowers flame-

red around December. Frothy white blossom on apple trees
produces orchard honey. The violet, snaky stalks of viper's bugloss
make a clear gold honey. Italian chestnut trees spread a dark
fragrance; mango honey is truly fruity.

Almost every flower can contribute some flavour – violet, caramel,
resin, butterscotch, raspberry. But bees are not snobs. Another observer
once saw an outlandish technicolor comb taken from a hive near a
sweet factory. The bees had been foraging on discarded seaside rock,
and the colours had come through in the honey.

The farming of honey – one can hardly call it the domestication of
the bee – seems to have begun in the Middle East, like so many agricultural practices. The first evidence of man-made hives is from a relief
in the sun temple of King Neuserre in Egypt, around 2500 BCE. The
hives look as if they are horizontal cylinders, stacked like an outsized
honeycomb. Later images show beekeepers holding pots which are
emitting stylized flames (and presumably pacifying smoke) towards the
hives. In these clearer representations the hives are grey in colour, perhaps made of mud from the Nile.

The scale of honey production in ancient Egypt was huge. A
papyrus records that Rameses III (c. 1170 BCE) provided 20,800 jars of
honey in a single offering to the Nile god. In 1978, the archaeologist
and honey expert Eva Crane counted 10,000 hives in a seventeen-mile
stretch of the Nile valley north of Assyut. The Egyptian beekeepers
were using almost identical mud-hives to those shown in the 3,000-
year-old reliefs.

The Greeks farmed honey, and believed that if bees hovered
near the mouth of a baby the child would grow up with the gift of

'honeyed words'. Virgil is said to have been so blessed, and celebrated bees in his poem on the craft of farming, the *Georgics*. The Romans made hives by threading together the dried stems of giant fennel, a technique that is still used in Sicily. In the forests of Europe in the Middle Ages, honey-hunters suspended hollow logs in trees for swarms to settle in, and later placed elaborately carved tree trunks on the ground.

In tropical America, the honey bees are not *Apis* species, but belong to the *Meliponidae* family. Unlike the European bees, these are stingless, but can be very aggressive, covering humans (including the insides of their nostrils and ears) in order to suck their sweat. Their diet is much more varied than European bees, including the sap of trees, bird-droppings and occasionally animal carcasses, and the honeys are consequently very different, in colour and consistency as well as chemical composition (their chief sugar is levulose, very much sweeter than sucrose). A few are toxic. Most are very dark, slow to crystallize, and highly flavoured. The anthropologist Claude Lévi-Strauss had first-hand experience of them in Amazonia, and writes with an unusual show of emotion that they have

> a richness and subtlety difficult to describe to those who have never tasted them, and indeed can seem almost unbearably exquisite in flavour. A delight more piercing than any normally afforded by taste or smell breaks down the boundaries of sensibility, and blurs its registration, so much so that the eater of honey wonders

whether he is savouring a delicacy or burning with the fire of love. These erotic overtones do not go unnoticed in the myths. On a more commonplace level, the high sugar content and powerful flavour of the varieties produced by the Meliponidae give honey a status which is not comparable with that of any other food.

A myth of the Ofaie-Chavante tribe in the Matto Grosso about the origin of honey is extraordinary in that it is almost the inverse of the myth about the origins of agriculture from the same region (pp. 74–5). In understanding this it's important to remember that in many pre-industrial societies honey is classified as a plant.

'In the beginning, the wolf was the master of honey, and though it smeared its children with honey every morning, it refused to give any to the other animals. But the tortoise was persistent in its demands, and in the end the wolf allowed it to lie on its back and drink honey from a hanging gourd. This was just a trick. The wolf piled wood around the tortoise and set fire to it, hoping to cook the animal in its shell. But it was the wolf who couldn't stand the heat, and fled. The animals gave chase, and many shape-shiftings and transformations later, they arrived at the honey's source, the bees' house – the entrance to which was guarded by poisonous wasps. A woodpecker, or a humming-bird, eventually succeeded in getting past the wasps and securing the honey. The tortoise distributed it among the animals, so that each one had a cutting that it could take home and plant.

'But the animals were greedy. When their leader inspected their plantations, he found that many had eaten the honey they'd been given to plant. "This can't last much longer," he said. "We shall soon be without honey." So he released the bees, which flew away into the forest. Later he called the animals together and told them to take their hatchets and set off in search of honey. "The forest is full of all kinds of honey: bora, mandaguari, jati, mandassia, cagafogo. All you have to do is go and look for it, and if you do not care for one particular kind of honey, you can move on to the next tree, where you will find another. You can collect as much as you like; the supply will never be exhausted, provided you only take away as much as you can carry in your gourds. But what you cannot take away must be left where it was for the next time, after you have carefully sealed up the opening you made." Since then, because of this, we have enough honey.'

What is unusual about this myth is that honey begins as a cultivated 'plant', which can be set in the ground to grow and ripen. The animals seem already to be in possession of the secrets of cultivation. But cultivated honey has disadvantages. It is too accessible, too tempting. The honey plants grow so well and are so easily harvested that over-consumption soon exhausts the supply.

What the myth suggests is that the transformation of cultivated honey into wild honey removes these disadvantages at a stroke. The bees disperse and diversify into different species. There will be many varieties of honey instead of one. The greed of the honey-gatherers will be limited by how much they can bring back. What is left will increase, ensuring a sustainable harvest.

'There is no doubt where the originality of this myth lies,' Claude Lévi-Strauss remarks about its extraordinary reversal of the usual flow of 'progress'. 'It is, one might say, "anti-neolithic" in outlook, and pleads in favour of an economy based on collecting and gathering, to which it attributes the same virtues of diversity, abundance and preservation claimed by most of the other myths for the reverse outlook, which is a consequence of humanity's adopting the arts of civilisation.'

One of the problems with this exotic and complex substance is that it's just too characterful. You cannot help but savour it. You are aware of the depths and subtleties of its aromas. It doesn't have the simple 'hit' of a dose of refined sugar. In the early 1800s, when the British consumption of sugar was a mere quarter of today's 53kg per head per year, there were already signs that its mildly addictive properties were beginning to have an impact. Sugar began to be added to everything – to meat and fish dishes and old-fashioned puddings and wine. But what finally sealed sugar's triumph over honey was the escalating popularity, across the classes, of another plantation crop: tea. Neither tea nor, to a lesser extent, coffee is enhanced by the unrestrained fragrance of honey, which tends to disguise the aromas of the drinks themselves.

This was convenient for the plantation owners, who zealously endeavoured to indulge this expanding public habit. But during the early nineteenth century the price of sugar plummeted, largely because the world market had become much more competitive. It was inevitable that a home-grown substitute for sugar-cane would be sought, and sugar-beet began to contribute much more to the national quota. Sugar-beet had been known as a natural sport of the native sea-beet since Neolithic times, but it was developed into a commercial crop in France and Germany in the nineteenth century. Napoleon supported the growing of the crop, as part of his boycott of cane sugar from the West Indies. It didn't really become significant in Britain until the 1920s, when the government encouraged its cultivation (and sugar consumption) as a palliative to the great agricultural depression of the inter-war years – an early example of the warping effects of feather-bedding in the farming industry.

The consequences of our modern sugar habit hardly need repeating. Obesity, heart disease and diabetes, in all of which sugar is implicated, are now worldwide epidemics. They are increasingly becoming diseases of children, too. Yet sugar continues to be one of the most heavily subsidized crops in Britain, and to be a major constituent of most processed foods. Which results in the bizarre situation of taxpayers underwriting sugar production with one hand and a national health service overburdened with its effects on the other.

The whole history of sugar has been a terrible parable of the conflict between intensive agriculture and human – and nature's – needs. From the moment sugar-cane was first spread east by the Arabs in the Middle Ages, sugar's narrative has been an almost unmitigated tale of slavery, exploitation, devastated landscapes and human disease. Few other plants – with the exception of tobacco – have been such an agency for harm in the world. Of course, in the poorer parts of the world, sugar growing (like tobacco cultivation) does provide employment, and a source of cheap energy. But the costs far outweigh these modest advantages. Some 7 million people are directly employed in the sugar industry, an economic benefit which is dwarfed by the multi-billion-pound costs of treating sugar-related diseases worldwide. And with 70 per cent of refined sugar now being consumed where it is grown, these diseases are spreading way beyond the developed countries. The only true beneficiaries of sugar-cane are the multinational corporations that promote its products and help dictate food policy across the planet.

None of this is the sugar-cane's 'fault'; but since we still use the word panacea for plants which humans believe to be unequivocally beneficial, perhaps we need one for plants whose human use has always been corrupting. 'Bane' might do. The word's current meaning is 'that which causes ruin, or is pernicious to well-being'. Up to the seventeenth century it was also used in plant names in combination with an animal, to indicate a specific poison, as in henbane, wolf's-bane, leopard's-bane. Sugar would well qualify as human's-bane.

Honey, of course, is a sugar too, though an unrefined one. But the quirkiness of taste that prevents it satisfying a society with an habitual need for an uncomplicated sweetener also ensures that it could never become a food of excess. Yet it's intriguing to think of the role it could have played in the modern food chain in the absence of cane sugar. Like other forest products, it emerges from what is essentially a three-dimensional farming system. It relies absolutely on the existence of other, complex habitats (flower-rich grasslands and woodland) and is labour-intensive at every stage in its harvesting and processing. It is, of course, still stolen from wild bees; and squaring that ethical dilemma is something most European beekeepers shy away from. Yet the honey-gatherers of north-west Zambia – with more than a wave of solidarity to their myth-making Amazonian cousins – are showing how it might be possible to farm (if not market) honey in an ecologically responsible way. The bee 'keepers' hang bark hives high in the trees, out of the reach of honey badgers and army ants. They're quickly occupied by wild swarms, and a couple of years later the bee-people return, climb the trees and crop the honey, leaving half for the bees. In 2003 alone, two thousand new Zambian honey-gatherers joined the cooperative association which has been formed to market their products. But they, of course, like the bees, are in another person's employ, and their produce is shipped expensively around the planet with the help of fossil fuels . . .

Leaves of grass

CONDITIONS ARE NOWHERE RIGHT AT EDEN for a surrogate field of cotton, the species that, after sugar, brought the ecological destructiveness and inhumanity of plantation farming to full fruition. But it does have a small field of hemp, *Cannabis sativa*, whose history as a fibre plant has challenged the assumptions of industrial farming for three hundred years. A thicket of its bright green, lupin-like leaves advances right up to the main pathways in high summer. It's a source of surprise and mischievous delight to visitors, because this is not a plant that you are allowed to see in the British countryside. The fact that the drugs marijuana and hashish derive from its leaves (though they are barely produced by the plant in the British climate) has made its cultivation illegal without a special licence. Or at least that is the official explanation.

Eden's hemp exhibit is a clever, teasing installation that challenges all the conventional clichés about the plant, including the assumption that its drug connections were what turned it into an outlaw. It is simply a fence – a fence made out of the very thing that's being fenced off. You see the hemp plants through a cat's-cradle of hemp strings. George Fairhurst, who made the piece, was a professional sailing-boat skipper for fifteen years, and knows his ropes. They are brisk and taut, stretched out straight for a while, then joining in Turks' heads and tangles, knotty problems. And along the fence they're also woven to form portholes into the plantation, some about the size of handcuffs, some big enough for your head. Visiting youngsters of course oblige, sensible of the fact that the most intelligent way to get past an obstacle is to use your head. A few of the older ones put their arms through too, snatch a bit of leaf and hide it hurriedly in their pocket. Squeezed and sniffed, it will give them imaginary highs for the rest of the day.

The captions for the exhibit list hemp's extraordinary history. Six thousand years at least in use as a fibre plant. A mainstay crop in medieval Europe for the production of clothing, sails, rigging, fishing nets. The material from which the first pair of Levis were made (and parts of the first Ford car). And now found to have an almost limitless list of uses not just for its fibres but for its oil too. But in the 1960s in Britain, and then throughout much of the Western world, production of this – perhaps the most versatile of all the world's crop plants – was shut down. The excuse, of course, was the spreading popularity of the mildly narcotic drugs produced by the plant. Even though these are barely present in the varieties grown for fibre, the cultivation of all kinds of hemp without a special licence was banned. But Eden's exhibit raises a legitimate question, even though the evidence is negligible. Is it credible that the influence of this mildly hypnotic drug was responsible for such *economic* strictures? Or had 'King Cotton' – as it had so often before – exerted pressure at a high level to suppress a crop whose future was just beginning to look competitive again?

I'm perhaps more sentimentally attached to hemp than some other enthusiasts. The house I live in was a small hemp farm up to the nineteenth century. In the earliest property-by-property map I've been able to find of the Waveney valley in south Norfolk (the 1839 Tithe Map), our farmhouse is shown with ten acres of 'Hempland' attached. The cannabis sprouted where our apples and plums now fruit. The eighteenth century was its heyday in this part of Britain, as one of the most popular non-food crops among smallholders and peasant farmers, and marked the climax of a story that had begun in Stone Age Asia.

Hemp is native to an area stretching roughly between Russia and Iran. It grows as an annual in damp, light soils, often reaching four metres tall, even in the wild. Its pale green, slightly limp, famously open-handed leaves give it an oddly foppish look for such a hefty plant. Cloth and rope made from the long fibres in the stalk were in use in western China five thousand years ago, as were, medicinally, the leaves and oil from the seeds. By 700 BCE the plant was being cultivated on the Russian steppes, and it seems to have reached Britain in the Anglo-Saxon period. Certainly it was being grown widely on monastic estates in east Norfolk at the beginning of the fourteenth century. It was a favourite crop among commoners in these parts, showing a high yield on soils not suitable for much else, and needing next to nothing in the way of manuring or weeding. During the European-wide inflation of the sixteenth century, when the prices of imported linen and canvas soared, and England's perilous reliance on a single export commodity – woollen cloth – was exposed, there was an official campaign to encourage the growing of alternative fibres. Henry VIII's ministers passed a law that everyone with arable land was to grow an acre of hemp or flax for every sixty acres of cereal, but it was never strictly enforced – if, indeed, it ever could have been.

Over the next three centuries the rise and fall of hemp's fortunes are a pointed lesson in how vulnerable non-staple crops can be to the swings and pressures of the market, regardless of their social desirability. In the sixteenth century hemp was all the rage, and as part of the campaign to increase the amount of home-grown and home-manufactured rope and cloth, the government encouraged local authorities to invite foreign weavers to settle in their towns. The Huguenots, already fleeing from religious persecution in northern Europe, needed little encouragement, and many were attracted to the non-conformist tradition of East Anglia. By 1600 there were increasing numbers of 'websters' (weavers) in the Waveney valley. The light, fenny soils suited hemp, and it became one of the commonest crops grown in gardens or in small strips in the open fields. The plots were rarely more than a few acres in extent. In his doggerel manual, *Five Hundreth Pointes of Good Husbandrie*, the East Anglian farmer Thomas Tusser suggests that the sowing and processing of hemp is woman's work ('Wife, pluck fro thy seed hemp, the fimble hemp clean'). But wills and registers of occupations suggest that the work was more evenly shared. What is certain is that in a large number of cases, the entire process from sowing to weaving was done within single families. Even where it wasn't, weavers were typically self-employed commoners, who in addition ran a cow and some geese on the common and tended a few fruit trees. The production of hemp cloth was the linchpin that held the local economy together.

Making the hemp ready for spinning and weaving was a complicated business, with its own patois. First the cut hemp had to be wettened – 'retted' – to separate the fibres from their woody rind. This was done either by leaving the stalks to lie for four to five weeks in the field where they'd been cut ('dew-retting'), or, more often, by steeping them in a pond for five days. Hemp retted in this way consistently fetched a higher price than the dew-moistened crops. (The retting-pond used at our hemp farm still survives.) Then the bundles of hemp were beaten with a wooden 'swingle' to break the rind, and 'scutched' with a long, flat piece of sharpened wood to remove it altogether. Finally, the fibres were 'heckled', straightened and combed until they were fit to spin into yarn.

By the early 1770s, one-tenth of all the hemp grown in England was produced in Norfolk, and most of this in villages within a ten-mile radius of our own farm. The range of craftspeople associated with the production of the cloth – hecklers, spinners, weavers, bleachers – was second in number only to general farmers. But this was the peak of hemp production. The war with France which resumed in 1778 made wheat a more valuable crop, and at the turn of the nineteenth century Arthur Young, Secretary of the Board of Agriculture, visited the 'vicinity of Diss', and found just two hundred acres of hempland left – scarcely one-tenth of what there had been thirty years before. This was also the great period of agricultural improvement, when aspirational landowners were bringing commons and fens into cultivation, amalgamating small farms into larger units and enclosing the open fields. As mainstream and essentially large-scale cereal agriculture moved into the ascendancy, encouraged by high prices and a growing population, so alternative crops and small-scale peasant systems retreated. It was a pattern that had been repeated many times since agriculture began in the Middle East.

Yet in two villages in the Waveney valley, the hemp trade persisted right into the latter half of the nineteenth century. The Lophams were referred to by a parliamentary commission on hand-loom weavers as 'one green oasis in the vast desert of discontent'. This was due to the efforts of a group of master weavers who organized themselves into a kind of cooperative, employing fifty other local weavers as outworkers.

They worked hard to sell their products, delivering them by donkey, and then, as outlets increased, by horse-drawn van. But they continued to produce hemp linen in the traditional way, working individually in their cottages, bleaching the cloth by laying it out on grass in the sun, and, in the words of the commission, entertaining 'a very primitive horror of the employment of chemical compounds' – a practice common in the north of England. Lopham linen, especially tablecloths and napkins, by now ornamented with the elaborate designs possible on Jacquard looms, became for a while the most famous in England. It went to the very best addresses – Eton and Harrow schools, Kensington Palace, the Russian Embassy – and received the Royal Warrant in Queen Victoria's reign.

Hemp growing and weaving had pretty well died out in East Anglia by the turn of the twentieth century, as they had done across the rest of England. Yet when there was a revival of interest in the crop in the 1960s, it was rapidly suppressed by legislation. Whether the cotton industry, sensing possible competition from cannabis grown for fibre, had a hand in the restrictions is unlikely ever to be known. But the new laws, which forbid the growing of hemp without special licence, have put an effective brake on the small-scale cultivation that was always part of the crop's history. The Home Office's criteria for issuing licences are exacting, and seem chiefly concerned to minimize the threat of *theft*, not drug abuse: 'It is for the grower to decide the most suitable location but *ensuring that it is only grown where there is minimal risk of attracting the attention of those who might steal the crop . . .* The following factors must be considered when selecting sites: *certain situations must be avoided – growing beside busy public roads or close to housing, industrial or leisure areas . . . certain factors, including topography, should be considered where, for example, folds in the land can be utilised to shield crops . . .'* They seem heavy-handed criteria for a crop with no noticeable narcotic properties, and where the plantings are big enough, the rules appear to be discreetly put to one side in the issuing of licences. The largest recent planting to return to south Norfolk, for instance, was several hundred acres just outside Thetford, clearly visible from the busy trunk road that ran beside the fields. The cannabis was being grown for the German firm BMW, who use its compressed fibres as a replacement for synthetic plastic in its car interiors.

The crop began its slow, licensed return in Britain in 1980, initially for making hempen paper (still the favourite for Bibles and cigarette papers). In 1992 an entire issue of the *Ecologist* was printed on the paper. In 1993 a cooperative of twenty East Anglian farmers began cultivating hemp over about 1,500 acres, also for paper, with the waste being sold for horse-bedding, as it is absorbent and makes good compost. (This last idea came from France, where more than 10,000 acres of hemp are currently under cultivation.) Potential outlets seem almost limitless, and include hemp as a fuel for power stations, as a building material, and as a blanket for soaking up oil from polluted water.

Yet there are still fewer than 5,000 acres licensed to hemp in the UK. Perhaps this will increase as varieties and harvesting technologies improve. Already there is a strain whose fibres mature while the stems are still soft and green, and which doesn't require long soaking to release them, and another which reaches a lower, more regular height, making harvesting by combine possible. We have to hope that these advances don't mean that hemp will become just one more mass-produced arable crop. Its high productivity and ability to thrive without artificial fertilizers and pesticides make it a perfect crop for smallholdings and organic farms, and it would greatly help its progress if the anachronistic ban on growing without licence were lifted.

'The chosen ornaments of royalty'

COTTON AND SUGAR-CANE, AND A HOST of other economically valuable crops, were all shifted about internationally in ways that were to permanently affect the economics and ecology of the planet. Cork oaks were dispatched to South Australia, tobacco to Natal, South American cassava to tropical Africa, Chinese tea to India. Kew Gardens (especially in its notorious hijacking and translocation of rubber from South America to south-east Asia) and private entrepreneurs like the East India Company coordinated and organized this activity, so that plants effectively became the shock troops of imperial expansion. Perhaps the most crucial tactic was the carrying of the cinchona tree – the source of quinine – from its native South America to south-east Asia and the Indian sub-continent. Without this source of an effective remedy for malaria, the European settlement of these areas would almost certainly have been short-lived.

Yet there was also a more subtle colonization of the tropical plant world, a corralling of its exotic wildness, a hunger to claim its beauty as a cultural adornment for the West. The medium for this domestication, the way both of magically transporting the flowers home and then displaying them, was a closed, transparent container – the prototype, both literally and metaphorically, of the biome.

Nathaniel Ward was a London surgeon and amateur naturalist, and early in 1830 he noticed that a young fern and a grass seedling had sprung up in a little moist soil he had accidentally left behind in a jar containing a hawk-moth pupa. Nothing unusual in that, perhaps –

except that the jar had been
sealed several months before.
Ward was struck with the
thought that this might be the
perfect way to grow lush and
uncontaminated plants in the
toxic atmosphere of the East
End. Not only would the plants
be protected from drastic temperature changes and noxious
fumes, but they could thrive
indefinitely without water. They
were, in effect, in a closed meta-
bolic loop. They exhaled carbon
dioxide by night, and absorbed it
again by day. The water which they
transpired condensed on the glass,
trickled down and was subsequently taken up again
by the plant's roots. With the help of a firm of commercial nurserymen, Ward had several of these
miniature greenhouses made up, and put them to the
test in his own house. They were highly successful, and news of these
'Wardian cases' soon spread through the learned and amateur natural
history societies that proliferated in Victorian England. What proved to
grow best in them were those lovers of continuous damp, the ferns,
and for a few years the British countryside was scoured for odd varieties and sports of native fern species. The popular writer Shirley
Hibberd enthused about the 'plumy green pets glistening with health

and beadings of warm dew' in their glazed cells.

But the fashion died, and naturalists and gardeners began to explore the potential of Wardian cases for exhibiting more exotic plants, in elaborate arrangements with shells and mounted butterflies. No fashionable drawing room was without its case full of 'vegetable jewellery' (Shirley Hibberd again). The cases themselves became increasingly extravagant. There were Oriental cases, Gothic cases, cases modelled on Tintern Abbey, even a few shaped prophetically as domes.

These kinds of collections for display in the home were nothing new. They could even, at a pinch, be placed in a lineage which went back to the accumulations of objects in prehistoric funerary hoards, simultaneously status symbols and a kind of charm to ensure a harmonious future life with nature. What were known as 'cabinets of curiosities', compartmented boxes full of gems, sponges, insects and fossils, had been popular since the seventeenth century. And the fashion for stuffed birds in glass cases had begun at least a century before Victoria. The unusual feature of Wardian cases was that they were living collections, and often assembled as tiny simulacra of the exotic locations where they were collected. It was as if the public had been given a way of sharing in the empire's prodigious new bounty (or booty), of bringing it home in the most literal sense.

Meanwhile the cases were being put at the service of the greater national project. The botanist John Loudon had seen their potential as soon as he set eyes on Ward's own collection. It was, he wrote, 'the most extraordinary city garden we have ever beheld ... The success attending Mr Ward's experiments opens up extensive views as to the application [of the cases] in transporting plants from one country to another; in preserving plants in rooms or towns; and in forming miniature gardens or conservatories.' But Ward had already realized the possible value of his cases in 'the conveyance of plants upon long voyages. Reflecting upon the causes of the failure attending such conveyance, arising chiefly from deficiency or redundancy of water, from the spray of the sea, or from the want of light in protecting from the spray, it was of course, evident that my new method offered a ready means of obviating these difficulties.'

So in 1833 two Wardian cases, filled like botanical arks with flowers and ferns, were sent on a great voyage to the Antipodes. They were lashed to the poop-deck for four months, and arrived in Sydney in perfect condition, and with a single primrose triumphantly in flower. This was practical stuff, but it's hard not to be reminded of the ancient archetype, the small island of life adrift on the waters, 'the gleaming membrane ... floating free'. For the return journey Ward had stipulated they should be filled with native Australian plants, including species which had already proved to be bad travellers. They were sent on a storm-racked journey round the Horn, and between Botany Bay and London the plants were subjected to temperatures which varied from −6.6° to 48.8° C / 20° to 120° F. But when Ward went to collect them from the quayside, he found 'the plants were all in beautiful health, and had grown to the full height of the case, the leaves pressing against the glass'. Soon the cases were in mass production, and in the 1840s were used to transport twenty thousand tea plants from Shanghai to the Himalayas for the East India Company.

The Wardian cases were increasingly used to move ornamental plants, and especially tropical orchids, a family whose exotic glamour gave them a powerful hold over the Victorian mind. The collector Frederick Boyle concluded that orchids were 'expressly designed to comfort the elect of human beings in this age'. Before the 1830s there were comparatively few tropical species in cultivation in Britain. The first to be successfully grown was a *Bletia* from the Bahamas, which was coaxed into flowering in 1731 in the glasshouse of Sir Charles Wager in Fulham. It died soon after. By 1789, there were fifteen species flowering at Kew. But the first truly to catch the imagination of the public – and plant collectors – was the luscious, pink-flowered *Cattleya labiata* from Brazil, a member of what is currently one of the most commercially popular orchid families. Its remarkable comings and goings may have contributed as much as its appearance to its glamorous reputation. It first appeared in the hothouse of William Cattley, whose collector, William Swainson, had sent him a cargo of plants from the Organ Mountains in Brazil. Out of curiosity Cattley had planted out the packing material from this collection, and from it sprang this prodigious bloom. For a while the *Cattleya* was the pride of exotic collections, but the specimens gradually died out until only one was left, which was then itself destroyed in a fire. By this time no one seemed sure of where the original had grown, and it became the Holy Grail of the orchid-collectors' world. The story of its subsequent rediscovery is possibly apocryphal. Seventy years after its disappearance, it re-emerged in Paris – rather improbably as the corsage of a woman at an embassy ball. An orchid enthusiast attached to the British Legation spotted it, and was sure he recognized it. An expert confirmed his guess, and a trail was followed back to the exact site near Pernambuco in Brazil where the plant had been gathered.

By the middle of the nineteenth century 'orchidomania' was in full swing. The richer owners of hothouses sent their agents and gardeners right across the tropics, to Guatemala, British Guiana, Mexico, Borneo and the Philippines. Extraordinary prices – sometimes up to a thousand guineas a root – began to be paid for the choicer plants. The star collector, the German Benedict Roezl, dispatched orchids by the ton, and his cargoes often contained in excess of one million plants. The damage to the orchids' native habitats was devastating. Forests were ransacked. Collectors routinely cut down thousands of mature trees to collect the epiphytic orchids living in their canopies – and then often eliminated a species from an area to thwart other collectors and maintain its rarity value. Forged maps were distributed to send rivals off on wild-goose chases. The director of the Zurich Botanical Garden said: 'This is no longer collecting. It is wanton robbery and I wonder that public opinion is not stronger against it.' When a specimen of the much-desired *Dendrobium schroederi* was offered for sale at auction in London, it had – to fulfil a promise made to the tribe who'd originally owned it – to be sold inside the human skull in which it grew.

Many of the orchids gathered so recklessly died, either on the voyage home or in the unsuitable hothouses they were consigned to. The ancient Western belief in the ubiquitous virtues of 'fertile' soil, and an ignorant and generalized view of the complexities of tropical habitats, meant they rarely got the growing conditions they needed. In the early days most were plunged indiscriminately into hot compost, and no distinction was made between orchids which grew in the ground and ephipytes, which were rooted to tree branches and survived on atmospheric moisture and trickles of rain. One of the early collectors, Conrad Loddiges, might have grasped the secret if he had attended to the success of an *Oncidium* from Uruguay, which had flowered repeat-edly on the voyage back to Britain, while hung up in his cabin without any earth.

Sir Joseph Dalton Hooker (appointed director of the Royal Botanic Gardens in 1865) collected orchids in the Himalayas in 1850, and his journals show the tunnel vision that even experienced botanists could display when confronted by these bewitching plants. In the Khasia Hills he found *Vanda caerulea* growing in profusion in woods of dwarf oak, and turned instantly into an orchid looter. This epiphytic orchid was exposed to 'fresh air and the winds of heaven ... winter's cold, summer's heat, and autumn's drought [and] waving its panicles of azure flowers in the wind' was growing in conditions completely at variance with the standard methods of cultivation in Britain, which penned all orchid species in hot, airless enclosures. Nonetheless, Hooker picked '360 panicles, each composed of from six to twenty-one broad pale-blue tasselated flowers, three and a half to four inches across'. They were to be preserved for 'botanical purposes', and made 'three piles on the floor of the verandah, each a yard high; – what would we not have given to have been able to transport a single panicle to a Chiswick fete!' Hooker adds a revealing footnote to this entry:

> We have collected seven men's loads of this superb plant for the
> Royal Gardens at Kew, but owing to unavoidable accidents and
> difficulties, few specimens reached England alive. A gentleman,
> who sent his gardener with us to be shown the locality, was more
> successful: he sent one man's load to England on commission, and
> though it arrived in a very poor state, it sold for £300 ... Had all

arrived alive, they would have cleared £1000. An active collector, with the facilities I possessed, might easily clear from £2000 to £3000 in one season, by the sale of Khasia orchids.

Meditating on why the lure of orchids surpasses that of all other plants, Eric Hansen (writing of modern 'orchid fever') is tempted towards a Freudian explanation: 'I took a closer look at the flower [a *Paphiopedilum* hybrid]. The shiny, candy-apple-red staminode that covered the reproductive organs was shaped like an extended tongue identical to the Rolling Stones logo. This shocking red protrusion nestled in the cleavage of two blushing petals then dropped down to lick the tip of an inverted pouch ...' Certainly the flowers of orchids are notorious mimics, and can resemble everything from miscellaneous human organs to spiders. But one gets a sense from the stories of the orchid frenzy that it was not simply the beauty and outlandishness of the flowers that mesmerized their devotees, but their origins.

These were the fabled air-plants, growing without visible nourishment in the most remote and beautiful regions of the earth. They were Edenic, flowers as they were before the Fall. To capture them was in one way to reclaim the Lost Garden. Writing of the tropical hothouses to which they were brought, the Victorian romantic novelist Charlotte M. Yonge was reminded of 'a picture in a dream. One could imagine it a fairy land, where no care, or grief, or weariness could come.'

Victorian glasshouses and conservatories had developed at much the same time as Wardian cases, and shared some of their symbolic power as shrines to the Empire. In their early days they were showcases for newly discovered South African heathers and Himalayan azaleas. But by the 1830s new methods of glass manufacture had transformed their design, and the new greenhouses, with their thin, uniform panes, were dramatically lighter and warmer. When coal- or coke-fired heating was added, it meant that the houses were capable of accommodating plants from the heart of the tropics.

The garden designer J. C. Loudon imagined conservatories 150 feet high, housing immense forest trees and, like the fantasy gardens of seventeenth-century Edenists, full of monkeys and parrots. In reality it was caged canaries and stuffed hummingbirds and an extravaganza of tropical desserts. Echoing an old Elizabethan practice, the mango and pineapple plants from the hothouse were wheeled in their tubs to the table, where the fruits could be picked straight off the bush by fashionable diners, just as if they were noble savages.

The ultimate expression of the Victorian orchid cult, the transformation of that complex yearning into what was both a ravishing work of art and a temple to Mammon, was James Bateman's massive book *Orchidaceae of Mexico and Guatemala*, published between 1837 and 1841. Bateman was the most eminent and fanatical of all Victorian collectors and growers, and by the 1840s his house at Knypersley had the most extensive collection in Britain. In the 1830s he conceived the idea of creating a lasting monument to them, in a book of extraordinary proportions and elegance. He described orchids as 'the chosen ornaments of royalty', and dedicated the volume to Queen Adelaide. One hundred and twenty-five copies were printed for subscribers, and each plate was reputed to have cost £200 to reproduce. The bookseller H. G. Bohn described it as 'the most splendid botanical work of the present age'. It was certainly, at twenty inches by thirty, the biggest botanical book of any age. In his introduction, Bateman sets the slightly decadent tone for what is to follow, when he describes the sole function of orchids as entertaining us with their fragrances and appearances, and providing 'a rich banquet in the temple of Flora'. What follows, in its solemn, tongue-in-cheek extravagance, is – intentionally or not – as much a satire on Victorian plant-worship as a reverent guide book.

The most conventional aspects of the book are its superb life-size paintings by two artists who are referred to simply as 'Miss Drake' and 'Mrs Withers'. As is the case with most early female flower artists, almost nothing is known about these, except that they lived in London and had some attachment, as 'Painters in Ordinary', to the royal family. But their paintings are among the finest of the time, and have a lustre and depth that catch something of the sheen of live orchids.

Alongside the pictures Bateman adds a commentary which is a bizarre mixture of cultivation tips, geographical anecdotes and running orchid jokes. Sometimes he dashes off an irrelevant quip to accompany a black and white vignette of a Mexican folk-dress, or of a cockroach which was found in a natural history cabinet, 'where it ate everything *except* a *Catasetum* orchid'. A few pages later there is another vignette of the underside of the cockroach, and beneath the heading 'The Tables Turned' runs the caption: 'There we had a portly, well-conditioned insect ... here we have an ascetic, half-starved wretch who might not have eaten an *Orchis* for a month.' Bateman's flights of fancy about the resemblances between orchid flowers and birds, frogs, monkeys (and monks) were transformed into surreal cartoons by J. Landells. Here is Bateman considering the similarity between the flowers of *Cycnoches* and swans: '*Cycnoches loddigesii*, perhaps, bears, on the whole, the closest resemblance to the feathered prototype; for the column (answering to the neck of the bird) is long and pleasingly curved, whereas that of *C. ventricosum* is lamentably short; the sepals and petals, too (wings) of the former are thrown wide open, which look better than to have them thrown entirely back as is the case with the latter . . .' But *C. ventricosum* is closest to 'the swelling bosom' of a swan, and if the two species were united, 'we should have a vegetable swan as perfect in all its parts as are the flies and bees with which the orchises of English meadows present us'.

The white–eyes' burden

WALKING IN THE HUMID TROPICS BIOME IS A MORE serious business than strolling in the Med. The pathways are dripping tunnels. Edging past the vegetation is like brushing through an endless series of damp bead curtains. We visitors follow each other in a thin convoy – a salutary reminder, maybe, of the way the first doughty explorers moved through this kind of environment. The plant labels nag at your memory. The ingredients on the back of spice bottles, maltreated house–plants, snatches from school geography books all float past. Manioc. Sisal. Iboga, a fierce hallucinogen. Patchouli, a sprawling herb, that I had no idea was related to our mints. Betel nuts, fruits of a small palm. Chewed in wads, they not only 'promote well–being' but turn 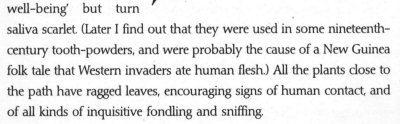 saliva scarlet. (Later I find out that they were used in some nineteenth-century tooth-powders, and were probably the cause of a New Guinea folk tale that Western invaders ate human flesh.) All the plants close to the path have ragged leaves, encouraging signs of human contact, and of all kinds of inquisitive fondling and sniffing.

I'm with Dina and her team, who look after the biome, and I'm curious about whether their principal aim – in keeping with Eden's defining theatrical ethos – is to give people an impressionistic flavour of the tropics, or whether they are also trying to give an authentic representation of rainforest vegetation. But no such simplistic purposes can really be clung to here. This is an environment fabricated, manipulated and compromised to a degree that dwarfs all Eden's other artifices. It's been assembled like an immense construction in organic Lego, and it is the performance of the end product rather than the integrity of the bits that matters.

Even the bedrock of the biome, the soil itself, is a clever piece of make-do. There is very little soil in rainforests, just a few centimetres of thin sand usually. The nutrients and energy are stored up in the canopy, and anything that falls to earth is recycled back into the system almost immediately. The trees prop up themselves – and each other – with complex, shallow root-braces. Eden began with no soil at all, not much space and no prospect of an internal community of recycling animals. There was very little air movement in the biome to rock the planted trees and encourage them to extend their roots. (At one point the team were thinking of a daily routine of shaking them by hand.) They had to be bedded down and supported in a way quite different from real rainforest denizens. Then there is the problem of the light. Eden is in a deep hole in the ground, and gets hours less daylight than surface structures. It also gets English winters – no real problem here in terms of temperature because of the artificial heating – but again dramatically reducing the amount of daylight tropical species are adapted to receive. The team have a dilemma here. If the heating is kept up sufficiently for the tropical plants' comfort, they will tend to grow too much foliage for the available light, and produce pallid and flabby leaves. Too little heat, on the other hand, could kill them off completely.

But this is nothing compared to the threat of pests and disease. These are Eden's nightmares. The workers in both biomes live in constant fear of contamination, of the penetration of their scrupulously virginal ecosystems, of their own version of the Fall. Extraneous life finds its way in, one way or another – as eggs in the root-balls of imported plants, or stuck to visitors' shoes, or just drifting in through the ventilation panels, as the odd dragonfly and bee do. Eden's freedom to retaliate is limited by the Project's own ecological principles, and heavy chemical attack is not an option, inside the biomes at least.

Dina points to their stands of young mangroves – a chance to show a slice of the rainforest's real roots for once, without having to justify them in terms of human usefulness. They're doing all right, but have been attacked by cockroaches. Upwardly mobile cockroaches. She has found them twenty feet above ground level, and doesn't know if they flew there or climbed up the trunks. There are already four different species, and Eden may yet aspire to hold the national collection of *Blatta*.

In the modest south-east Asian vegetable garden, there are spider mites on the tomatoes and tropical white-footed ants tightroping along the washing line. The beans are being eaten by scale-insects – extraordinary creatures, almost all female and lacking both wings or legs, but prodigiously fertile. They can kill whole trees. Dina looks at this teeming menagerie as if they were gatecrashers at her first childhood tea party. She does not want them spoiling things, but does not know how to make them go. Biological control is the first line of defence, and a few pairs of white-eyes have been introduced from Malaysia. These delicate birds, similar to Old World warblers in appearance, swoop about the canopy, and can be seen earning their keep, snapping insects off leaves. They seem to like the place, and are already breeding, making exquisite nests from fibres picked off the coconut palms. There are a few lizards too, but that is as far as introduced animals will go. Eden is adamant that it will not become a zoo, or even a butterfly park. No animal will be introduced here unless it serves some utilitarian purpose in keeping other animals at bay.

A gale is getting up beyond the shelter of the biome, and I'm suddenly struck by the familiarity of this argument. Whether wild animals had a place in a reconstituted Paradise has troubled thinkers ever since the Eden myth gripped the Western imagination. There are no animals at all in Dante's description of the Garden of Eden in *The Divine Comedy*. He describes being attacked by a lion, a spotted panther and a she-wolf at the beginning of the book, as he emerges from the depths of the forest, and plainly regarded their savagery – acquired at the Fall – as irredeemable. Yet by the seventeenth century, Christian idealists who believed that they could somehow 'reassemble' Eden in parks and

gardens began to admit animals as well as plants to the cast list – at least in theory. Although there were exotic creatures kept in the botanic garden at Uppsala, most gardeners realized the impracticality of admitting lions to the ordered beds. Most settled for a few stuffed beasts, or their images tidily clipped in box. But artists were not so constrained. G. B. Andreini's *L'Adamo, sacra rapresentatione* (1617) shows a formal botanic garden, with regular beds and paths, and an assortment of animals strolling about among the parterres, including what appear to be a monkey, a hyena, a marten, two rabbits, a chicken and a small snake. They look very well-behaved.

The one class of creatures about whose desirability all parties were agreed was birds. Du Bartas's imagined Eden had a thousand different sorts. Milton's Paradise was full of them, and John Evelyn wanted to construct an aviary in his botanic garden large enough for five hundred linnets, larks and thrushes. Yet it wasn't their ability to keep down the numbers of spider mites and white-footed ants that was admired so much as their songs. 'The Animal World does not afford more agreeable Objects to the Eyes, nor none that so sweetly gratifies the Sense of Hearing', wrote the artist Eleazar Albin in 1737. 'They were undoubtedly designed by the Great Author of Nature, on Purpose to entertain and delight Mankind, who for their Generosity are well pleased with these pretty innocent creatures'. The biomes' robins and white-eyes suggest that this sentiment is still strong.

Later, a less sweetly gratifying noise begins to rise in the biome. The wind has turned into a tempest, and is hurling sheets of diagonal rain at the acrylic covers. As the temperature outside plummets, the hexagons contract and begin to throb, like immense diaphragms. The noise is stupendous, as if a jetliner was hovering just overhead. I was once in a medieval church in Suffolk, listening to a performance of Benjamin Britten's *Noye's Fludde*, at the moment a great storm broke. The thick stone walls seemed barely adequate to keep it out. The choir sang 'For those in peril on the sea', and the terrified children crept about the stage with waves made of blue streamers. Walking on through the now-thunderous biome, this new Eden feels like an Ark as well as a Garden.

Yesterday, a huge storm on the north Cornwall coast had swollen the rivers on Bodmin Moor, and sent a surge of water downstream that had half destroyed the village of Boscastle. It was on exactly the same date in August as the Lynmouth flood disaster half a century before, yet the gossip is all about global warming as the culprit.

In the tropics storms are the fulcrum of natural change. Edward O. Wilson's account of a 'Storm over the Amazon' is one of the great evocations of the energy and resilience of forest ecosystems, and an alternative view of disruptive forces.

> In the midst of the chaos something to the side caught my
> attention. The lightning bolts were acting like strobe flashes to
> illuminate the wall of the rain forest. At intervals I glimpsed the
> storied structure: top canopy 30 meters off the ground, middle
> trees spread raggedly below that, and a lowermost scattering of
> shrubs and small trees. The forest was framed for a few moments
> in this theatrical setting. Its image turned surreal, projected into the
> unbounded wilderness of the human imagination, thrown back in
> time 10,000 years. Somewhere close I knew spar-nosed bats flew

through the tree crowns in search of fruit, palm vipers coiled in ambush in the roots of orchids, jaguars walked the river's edge; around them eight hundred species of trees stood, more than are native to all of North America; and a thousand species of butterflies, 6 per cent of the entire world fauna, waited for the dawn ... that is the way of the non-human world. The greatest powers of the physical environment slam into the resilient forces of life, and nothing much happens. For a very long time, 150 million years, the species within the rain forest evolved to absorb precisely this form and magnitude of violence of nature's storms in the letters of their genes ...

A gust of wind whips through or lightning strikes the tree trunk, and the limb breaks and plummets down, clearing a path through to the ground. Elsewhere the crown of a giant tree emergent above the rest catches the wind and the tree sways above the rain-soaked soil. The shallow roots cannot hold, and the entire tree keels over. Its trunk and canopy arc downward like a blunt axe, shearing through the smaller trees and burying understorey bushes and herbs. Thick lianas coiled through the limbs are pulled along. Those that stretch to other trees act as hawsers to drag down still more vegetation. The massive root system heaves up to create an instant mound of bare soil ...

I wonder if it's this sense of mayhem and renewal that I'm missing here, that's responsible for a nagging feeling that things are too predictable, too orderly. I watch the children, amazed at how respectful and well-behaved they are. Many of them duck under the mist-sprays for a brief relief from the heat, but not one dashes into the undergrowth, or tries to shin up one of the very climbable trees. It makes me a little wistful, though I know that Eden (and most certainly its insurance company) could never tolerate such behaviour, any more than it could allow unstable trees to fall of their own accord, as they do so thrillingly in Wilson's account of the Amazon. When a thirty-foot tall *Albizia* did come down suddenly in this cramped space, it almost took one of the gardeners with it. (But then this is a notoriously impish plant. A large clutch of *Albizia julibrissin* seeds – 147 to be precise – sprang to life on their herbarium sheets in London's Natural History Museum after it was bombed in the Second World War.)

Yet without some experience of the energy of the forest, some chance of witnessing that it is a living system, not just a still-life, we have only half understood its plants. And maybe worse, because what we witness is a scene entirely composed and controlled by humans. Without an appreciation of the role and contribution of wildness on the planet, of plants' own adaptability and inventiveness when the powers of the physical environment 'slam' into them, we are missing a crucial element of how they work.

I wonder for a whimsical moment what would happen if Eden abandoned the biome altogether, sealed it up like a Wardian case and left it to its own devices. Would the scale-insects proliferate and eat everything to death? Would the white-eye populations soar, and then plummet from starvation? Assuming the acrylic shell held, what would become of the big forest trees, as they packed ever closer together, their tips beginning to grow downwards instead of up? Would the system end up like a thrown-away goldfish bowl, full of a few species of aggressive algae and weeds? Already that indomitable local plant, furze, has found its way into the biome, growing far beyond human

159

reach on some of the sheer rock faces. Might an abandoned biome, by some deep eco-logic, drift towards the condition of the world just outside?

When they tried the experiment, in Biosphere 2 in Arizona – but with humans in the dome as well as plants – it ended in a kind of catastrophe, at least in human terms. The crops failed, the humans fell ill, and the internal ecosystem, such as it was, reverted to a crude, simplistic form suited to the conditions. The planners had underestimated, by several degrees of order, the complexity of viable ecosystems, the contributions made by myriads of small and anonymous organisms, the absolute necessity for an immense diversity of life forms to give the system flexibility and resilience. They had built a model, and expected it to work like the real thing.

Eden has never pretended that either the Project itself or the biomes are self-contained systems, or indeed that they are ecosystems at all, in the proper sense of the word. But it must find a way of *showing* this to its visitors, if the importance of the deep, encoded wisdom of wild growth is to be understood.

There is a buzz in the biome. The word is out that Sir Ghillean Prance, legendary Amazonian botanist, ex-director of Kew Gardens and a mentor of the plantings in the tropical biome, is paying a visit. He has come to inspect the cassava exhibit, and pass judgement on its authenticity. As he walks up the slope, a throng of eager acolytes form a wake behind him, like junior doctors round a great consultant. Sir Ghillean is a man of impossible distinguishedness, tall, upright, slow of step and resonant of voice. His eyes are constantly scanning – somewhere, I guess, around what corresponds to the lower levels of a rainforest canopy. He pauses before the cassava, and scans this too. Cassava, or manioc, is one of the most important staples in the wet tropics (and was served to generations of unwilling British schoolchildren as tapioca), but its roots are toxic in an untreated state, and need to be crushed and repeatedly washed before they are safe to eat. The Eden team have arranged the props of this process like a shop-window display, beside some planted cuttings of the shrub: a canoe, a few implements for slicing and grinding, a kind of frame on which the sacks of washed roots are hung to drain.

But the professor is shaking his head. There is only half the story here. The equipment is inadequate, and no distinction has been made between different styles of processing. He goes carefully through the entire routine, improvising the shape of a new display as he goes. The roots should be soaked inside the canoe, and turned with its paddle. Then they have to be baked – a stage entirely missing from the current exhibit. In the Amazon this is done on a large metal plate set in clay, about the size – Sir Ghillean gestures with his hands – 'of an Amazonian water-lily pad' (the local name for both objects is the same). He suggests that a local blacksmith should be commissioned to knock one up. Except that it is only the roots that are intended to make meal – *farinha* – that are baked in this way. The starch that is intended for tapioca is not baked, and is sometimes sent out of the settlements to factories. It is beginning to sound complicated, and those responsible for the display are furiously scribbling notes. Why not explanatory photos? Sir Ghillean suggests. But Eden is against such second-hand images on principle. So perhaps a full-scale facsimile of a *casa di farinha*, the mud huts in which the grinding and straining of the roots is done? 'It could be a natural gallery. The canoe could stand outside.'

Sir Ghillean returns to scanning the canopy. For a moment, standing on the raised bed, he looks like an Old Testament prophet. He is concerned about the composition of the vegetation. At present it is largely fast-growing, secondary forest species, the kind that shoot back quickly after a rainforest has been felled. This was fine to establish the display, but it's time to begin introducing more of the ancient, primary species. He spots one in flower, one of a large group of evergreen trees that are among the most primitive in the forest. This one, he announces, is pollinated by bats. The slightest of winces passes across some of the teams' faces at the vision of such creatures loose in the biome.

Warming to his guru's role, Sir Ghillean launches on another pollination story. The Amazonian water lily, whose pads provided him with a highly graphic and appropriate model for the cassava baking tray, and which grows in a pool not many yards from where we're gathered, goes through an extraordinary process of fertilization. In the tropics, the flowers open at night. They have an intense scent and are 11° C / 51.8° F hotter than the air outside. The warmth and smell attract beetles, which enter the elaborate structures of the flower. The flower then shuts, trapping the beetle for twenty-four hours while it does its pollination duties. Then it releases the beetle, to do its good works elsewhere. He has seen it happen himself, on the Amazon.

Vegetable wonders

I'D GO TO SIT BY THE AMAZONIAN WATER LILIES most afternoons. Their floating leaves – vast, formal, neatly crimped at the edges – looked more like pre-cast water features than anything alive. And each time I was there the flowers, maddeningly, were either well opened or tight-budded. I'd hoped to watch them – or rather smell them – unfurl. When this prodigious plant was first exhibited in London in 1850, tens of thousands of people made the trip to Kew Gardens to experience the full pomp and sensuality of the blooming of this 'vegetable wonder'. Kew's specimens were more punctual time-keepers than Eden's. At about two o'clock each afternoon, the new white buds (they are the size of tennis balls) began to emit a strong aroma, variously compared to melon, strawberry and pineapple. A few hours later the petals opened and began to change colour to rose-pink. Towards ten in the evening they started to close. The flower's slow decline continued the following day, when the fading petals became 'a drapery of Tyrian purple' until they finally sank beneath the water.

The story of the discovery of the water lily and the race to bring it to bloom in a British glasshouse, of its triumphant reception at Kew, and finally of how the Crystal Palace was inspired by the structure of its ribbed leaves, is a perfect parable for the Victorians' acquisitive, aspirational love affair with plants, of their slightly self-congratulatory glory in 'vegetable wonders'. Even the first telling of the water lily's story has an authentic nineteenth-century flavour. In September 1837, the members of the Society of Practical Botanists met at a pub in the Strand to hear a paper by a young Silesian explorer, Robert

Schomburgk, who'd been sent by the Royal Geographical Society to British Guiana. This is what he had found:

> It was on the 1st of January 1837, while contending with the difficulties that nature interposed in different forms, to stem our progress up the River Berbice, that we arrived at a part where the river expanded and formed a currentless basin. Some object on the southern extremity of this basin attracted my attention, and ... animating the crew to increase the rate of their paddling, we soon came opposite the object which had raised my curiosity, and, behold, a vegetable wonder! All calamities were forgotten; I was a botanist, and felt myself rewarded! There were gigantic leaves, five to six feet across, flat, with a broad rim, lighter green above and vivid crimson below, floating upon the water; while, in character with the wonderful foliage, I saw luxuriant flowers, each consisting of numerous petals, passing, in alternate tints, from pure white to rose and pink. The smooth water was covered with the blossoms, and as I rowed from one to the other, I always found something new to admire. The flower-stalk is an inch thick near the calyx and studded with elastic prickles, about three-quarters of an inch long. When expanded, the four-leaved calyx measures a foot in diameter, but is concealed by the expansion of the hundred-petaled corolla.

Schomburgk sent seeds of the water lily to Kew, and asked for it to be named *Nymphaea Victoria*, after the young Queen, who had agreed. But the fate of the seeds was rapidly overtaken by a melodramatic dispute – touched with Victorian sanctimoniousness and moments of farce – over the propriety of the new plant's name. Closer examination had shown it not to be a *Nymphaea* at all, but a member of a new genus, and the President of the Royal Society proposed it should instead be named, with full pomp, *Victoria Regina* (though, in a wonderful muddle, it was, in different publications, simultaneously referred to as *V. reginae*, *V. regalis* and *V. regia*).

But there was worse to come. The shocking discovery was made that the Queen's water lily had already been found and described by several earlier foreign botanists. One of them had given it the decidedly unregal name of *Euryale amazonica* back in 1832, and by the strict rules of botanical nomenclature (first named is best named) this was what it must be called. Nobody dared tell the Queen.

But again the taxonomists had made a blunder. *Euryale ferox*, from the East Indies, is another tropical lily of generous proportions. But 'the Queen of Lilies' turned out not to be from its genus. Protocol now ruled that the first successful bid for each half of the name should stand, so the plant became officially *Victoria amazonica*. The Queen's name had been restored, but with a rather compromising epithet in attendance. The director of Kew Gardens, Sir William Hooker, gravely pronounced that the specific suffix was 'totally unsuited to be in connection with the name of Her Most Gracious Majesty and must therefore forthwith be rejected'. Etiquette, perhaps for the only time in botanical history, triumphed over science and during the Queen's lifetime the water lily was always referred to as *Victoria regia*.

But all this was merely a sideshow compared to the spectacular piece of theatre in progress in the vivid light of the nation's greenhouses. Schomburgk's first batch of seed had failed to germinate, and his second (dispatched, on Hooker's advice, in phials of spring water) was divided between Kew, the Duke of Northumberland at Syon, and the Duke of Devonshire at Chatsworth. The gardener at Chatsworth was none other than Joseph Paxton, who lovingly nursed the seeds in a specially constructed tank kept at 29.4° C / 85° F. Each of the gardens planted out the seeds in conditions they thought most likely to produce the first flower, and the contest turned into a race 'as exciting in its day', Wilfrid Blunt has written, 'as Scott's and Amundsen's to the Pole, or the Americans and Russians to the moon'.

The plants at Kew and Syon proved incorrigibly shy, but the one in the Great Conservatory at Chatsworth grew so aggressively that its tank had to be enlarged. And on 2 November 1849, Joseph Paxton was able to write excitedly to the duke (then in Ireland): 'Victoria has shown a flower!! An enormous bud like a poppy head made its appearance yesterday. It looks like a large peach placed in a cup. No words can describe the grandeur and beauty of the plant.' The duke rushed home, and had a flower and a leaf sent to the Queen at Windsor.

Kew's giant water lily bloomed the following summer, and continued its display right up until Christmas. It proved one of the Gardens' greatest attractions to the plant-infatuated citizens of London, and the director had a much larger tank built to show it off to best advantage. Alas, as so often at this time, European horticultural reflexes proved quite inappropriate to the raising of tropical plants, and *Victoria*, deprived of decent ventilation and flexible heating, had to be moved again.

But it was the leaves more than the flowers that were to prove the enduring wonder of *Victoria amazonica*. They'd been likened by one observer to immense floating tea-trays, and the ribbing of the undersurface to 'some strange fabric of cast iron, just taken from the furnace'.

Several expensive folios of paintings of the water lily were published, including one by Walter Fitch which had lithographs reduced from originals twenty feet across. Designers used the leaves' distinctive shape and ribbing, incorporating them into chandeliers and gas brackets. One marketed a papier-mâché *Victoria* cradle.

In British Guiana, travellers had also noted that the Indians used to rest their children on goatskins on the floating leaves while they were working, and later in November 1849 the experiment was repeated with great success at Chatsworth. The Duke of Devonshire and Lady Newburgh placed Joseph Paxton's seven-year-old daughter, dressed as a fairy, on one of their home-grown leaves:

> On unbent leaf in fairy guise,
> Reflected in the water,
> Beloved, admired by hearts and eyes,
> Stands Annie, Paxton's daughter.

Meanwhile, Annie's father's skill in designing glasshouses had lifted him above his earlier station as a gardener, and in the 1850s he was invited to design the Crystal Palace. *Victoria* was his inspiration. 'Nature', he wrote, ' has provided the leaf with longitudinal and transverse girders and supports that I, borrowing from it, have adopted in this building'.

And in June 1889, Kew was host to a plant perhaps even more memorable than *Victoria amazonica*, an aroid which was described as 'one of the sensations of the London season'. In 1878, in the mountains of

Sumatra, the Italian botanist Odoardo Beccari came across a plant of gigantic and scabrous proportions. It is best visualized as an enormously bloated version of our own lords and ladies, complete with a crimson sheath and foetid spike. Beccari wrote:

> The single flower, or more correctly inflorescence and the tuber (from which it springs almost directly), form together so ponderous a mass, that for the purpose of transporting it, it had to be lashed to a long pole, the ends of which were placed on the shoulders of two men. To give an idea of the size of this gigantic flower, it is enough to say that a man standing upright can barely reach the top of the spadix with his hand, which occupies the centre of the flower, and that with open arms he can scarcely reach half way round the circumference of the funnel-shaped spathe from the bottom of which the spadix arises.

Beccari was able to take seeds of the aroid, by now admiringly christened *Amorphophallus titanum*, back to Italy, where they were grown on in the garden of his friend the Marchese Corsi Salviati in Florence. The Marchese presented some of the young plants to Kew, along with an altogether more extraordinary offspring: a life-size painting of the Titan, on a canvas eighteen feet by fifteen, showing a leaf growing out of the ground, and two Sumatrans carrying a tumescent inflorescence lashed to a pole. For a while the picture adorned the roof of the Orangery at Kew until Victorian prudishness banished it to a less brazen position.

But no such delicate sensibilities were evident towards the plants themselves. They took ten years to reach the flowering stage, but their growth in the last two was prodigious. In 1887 the tuber was three feet nine inches in circumference, and the leaf six feet tall. By 1889 the 'flower-bud' was growing at the rate of three inches a day. The Titan's flower finally opened at 5 p.m. on 21 June, and closed again at 11 a.m. the following morning. But its unfurling had been well advertised, and enormous crowds gathered to witness its brief moment of glory. What Beccari had failed to mention was its overpowering stench, described by one observer as 'suggesting a mixture of rotten fish and burnt sugar'. (Beccari himself compared it to 'ten thousand dead elephants'.) The crowd reacted with fascinated horror at this awful vegetable and its gross 'structure of recollection', and one of Kew's artists, Matilda Smith, was solemnly thanked by the editors of the *Botanical Magazine* for her 'prolonged martyrdom' in attempting to draw it. When she had to repeat the process with another of Dr Beccari's stinkbombs, the orchid *Bulbophyllum beccari*, 'which rendered the tropical Orchid house at Kew unendurable during its flowering', she was taken ill and had to abandon the project.

Throughout the nineteenth century, Kew and other gardens continued to put on displays of spectacular new plants, including giant cacti and the orchid *Cattleya skinneri*, which was six feet high and bore 1,500 flowers. Public interest seemed insatiable. But to say that this was a consequence of the national obsession with botany – and, for that matter, with all branches of natural history – is simply rephrasing the phenomenon. It seemed not just the intrinsic appeal of these vegetable wonders that drew the public, but the fact that they were being brought back to *Britain*. They were trophies, confirmation of the growth and legitimacy of the empire. And the more extraordinary they were, the more they seemed a benediction on the great enterprise. If the giant water lily was fit for a queen, then quinine and rubber were surely blessings on the nation.

Even less exotic species were appropriated for the great plant carnival, provided they were physically impressive. Quite early in the nineteenth century, an enormous hogweed was discovered by plant collectors in the Caucasus mountains. It grew up to fifteen feet tall, and had flower-heads the size of cartwheels. It was named *Heracleum asperum*, the 'Siberian Cow Parsnep', and on its arrival in Britain it was rapidly promoted as the temperate zone's answer to the flamboyant monsters of the tropics. John Loudon, one of the arbiters of Victorian gardening taste, praised it in *The Gardener's Magazine*:

> Its seeds are now ripe; and we intend to distribute them to our friends: not because the plant is useful, for we do not know any use to which it can be applied; but because it is extremely interesting from the rapidity of its growth, and the great size which it attains in five months ... We do not know a more suitable plant for the retired corner of a churchyard, or a glade in a wood; and we have, accordingly, given one friend, who is making a tour in the north of England, and another, who is gone to Norway, seeds for depositing in proper places.

By 1849 it was being offered for sale commercially by Hardy and Sons of Maldon, Essex, as '*Heracleum giganteum*, One of the most magnificent Plants in the World'. In 1870 William Robinson recommended them as highly suitable for any rough places near water, especially 'where bold foliage may be required', but added, prophetically, 'when established they often sow themselves, so that seedling plants may be picked up in abundance near them; but it is important not to let them become giant weeds.' Needless to say, they did break out, and we now know them as giant hogweed, one of our three most vilified 'noxious weeds' – illegal to plant outside gardens, and responsible for pernicious skin rashes on children who handle their stems in sunny weather. Of course, in flower, they seem as dramatically seductive as they did in Victorian times, and here and there, on a motorway verge or a river-bank or a damp industrial back-lot, you may glimpse a sheaf of these vast, angular parasols like some botanical henge, a reminder of an ear-lier generation's plant worship.

Visions of Eden

URING THE GREAT PERIOD OF PLANT COLLECTING, the botanic gardens also accumulated plant paintings. At Kew there are a million or so, filed among the even greater numbers of dried and mounted plants in the herbarium. I spent two months browsing through them in the late 1980s, a glimpse of an alternative approach to 'capturing' plants.

At that time the herbarium had sick building syndrome. The file of users' complaints was thicker than any of the plant folders. I had to abandon work at about three o'clock every afternoon, half blind and headache-ridden. I wondered if it was some kind of retribution by the plants, a delayed sting (like the pressed nettle which nipped the curator of the Linnean Society's herbarium two centuries after it was picked), or a last musty echo of their living odours. The herbarium specimens were far from chemically dead. Mineral prospectors would sometimes come and scour through the sheets, and take tiny snippings from species that were known to absorb metals from the soil. They would analyse these to judge their metal levels, and, if they were lucky, be able to identify the site at which the plant grew from the collector's notes on the sheet. The next stage would be the exploratory drills. The plants on their paper sheets had become compasses, storehouses of potent (and sometimes devastating) geographical information.

The paintings also seemed full of cryptic narratives. They were of every conceivable kind: sketches made on field trips, elaborate studio portraits, collections of colonial wives' watercolours, meticulous scientific diagrams. Yet even the most mundane collector's illustration conveyed something about how and when and where it was made. Like the Eden Project's artworks they often appeared inclusive and questioning, in a way that dried plants and their accompanying labels rarely did.

The European explorers and plant collectors of the eighteenth and nineteenth century weren't just annexing landscapes and appropriating whole ecosystems in the interest of imperial expansion. They were also colonizing nature itself, abstracting it from its independent life and its intimate connections with indigenous cultures, and reframing it in the assumptions of Western science. Mary Louise Pratt has written: 'One by one the plant's life forms were to be drawn out of the tangled threads of their life surroundings and rewoven into European-based patterns of global unity and order. The (lettered, male, European) eye that held the system could familiarise ('naturalise') new sites/sights immediately upon contact, by incorporating them into the language of the system.' The picturing of plants during the same period played its role in this project – as did the journals that many of the painters kept. Yet they often have a frankness and open-mindedness, too, an intrinsic respectfulness. Collectively they give an alternative view of the character – and status, if you like – of the vegetation of Europe's imperial estate. This is partly because of the nature of painting itself: the long conversation with a living plant that representation necessitates creates its own kind of intimacy. But it was also a consequence of the kind of people the painters were. Many were from quite different social and cultural backgrounds from the scientists – young self-taught gardeners, for instance, and self-employed women – and they had their own perspective on the West's grand project of, in Bacon's words, 'enlarging the bounds of human empire'. Their views of the New World's Edens inclined towards leaving them where and as they were.

Britain's first official roving plant collector and illustrator was a young Scot called Francis Masson. He was born in Aberdeen in 1741 and, like many of his talented countrymen, had to move south to find work. He got a job as an under-gardener at Kew, where his enthusiasm for botany soon attracted the attention of Sir Joseph Banks. Ever since his tantalizing first visit to the Cape in 1771, Banks had been eager to send an envoy there, to gather seeds and plants for Kew. A professional gardener, with an eye for attractive species, would fit the bill perfectly, and Masson was chosen (or volunteered) for the post. Perhaps neither man knew at that point that Masson had a natural talent for flower painting, and a latent sympathy for the cultures and ecosystems he was about to visit, and help reveal to the world.

Masson set out on his exploration from Cape Town in December 1772, and over the next two years, in addition to his drawings, he kept a methodical and revealing journal of his travels. He records the crops grown by the Dutch farmers, and the wild animals of the plains (he saw lions, elephants and zebras). His feelings about these are untypical for the eighteenth century, in the concern he shows for their future, and he laments the fact that the hippopotamus, once abundant in all the large rivers, was now almost extinct. 'This tract of country has afforded more riches for the naturalist than perhaps any other part of the globe. When the Europeans first settled there, the whole might have been compared to a great park, furnished with a wonderful variety of animals ... but since the country has been inhabited by Europeans, most of these have been destroyed or driven away.'

But at least the flora seemed inexhaustible. Masson began sending back flowers which were to become favourites in European gardens: ixias, gladioli, pelargoniums, mesembryanthemums and heathers. Yet it was the dry and seemingly barren sands of the western coast, known as the Karro, that intrigued him most. This is South Africa's Mediterranean zone, and he was fascinated to find that its plants, unhampered by the usual landowning and agricultural obsession with the fertility of the soil, managed to thrive in impressively arid and 'dismal' conditions. The succulents, especially, seemed 'endowed by nature, as the camel is, with the power of retaining water'. Among these were the strange and almost unknown *Stapeliae* which were to form the subject of his finest paintings. His journal states:

> At night we got clear of the mountains but entered a rugged country, which the new inhabitants name Canaan's Land; though it might be called the Land of Sorrow; for no land could exhibit a more wasteful prospect; the plains consisting of nothing but rotten rock, intermixed with a little red loam in the interstices, which supported a variety of scrubby bushes, in their nature evergreen but, by the scorching heat of the sun, stripped almost of all their leaves. Yet notwithstanding the disagreeable aspect of this tract, we enriched our collection by a variety of succulent plants, which we had never seen before, and which appeared to us like a new creation ... [A 'peasant'] told us, that in winter the hills were painted with all kinds of colours; and said, it grieved him often, that no person of knowledge in botany had ever had an opportunity of seeing his country in the flowery season. We expressed great surprise at seeing such large flocks of sheep as he was possessed

of subsist in such a desert; on which he observed, that their sheep never ate grass, only succulent plants, and all sorts of shrubs, many of which were aromatic, and gave their flesh an excellent flavour.

The *Stapeliae* are not cactuses but are fleshy and spiny enough to pass for them. There are about a hundred species, largely confined to Africa, and they share one striking feature: exquisite star-shaped flowers, with an incongruously strong and nauseating smell of rotting meat. It was these blooms that Masson captured so well. He drew a few wild specimens on his field trips, but based most of his work on plants he grew in his garden in Cape Town, where he'd settled in 1786. 'The figures,' he wrote, 'were drawn in their native climate, and though they have little boast in point of art, they possibly exhibit the natural appearance of the plants they represent, better than figures made from

subjects growing in exotic houses can do'. Masson was being typically modest. His collection of paintings, published as *Stapelia Novae* in 1796–7, catch the wonder and optimism he felt at this blooming of the desert. He had discovered the fecundity of the South African *garrigue*. The flowers themselves have the look of exotic sea-creatures, clinging precariously to the stems, or fragments of brilliant mineral drawn up from that dismal and 'rotten rock'.

Occasionally, attempts were made to bring botanical art itself into the service of empire, though not usually with much success. A unique venture was the project launched by the East India Company to use indigenous artists in logging the stupendous botanical riches of that barely surveyed sub-continent. In the early nineteenth century, one company employee became quite breathless about the prospect: 'What a vast field lies open to the botanist in that boundless country. How many unemployed individuals are there whose leisure hours might be agreeably usefully and profitably employed in this pursuit! ... Great God, how wonderful, how manifold are Thy works!'

The company had come to India in the early seventeenth century, and had acquired a remarkable degree of power for what was notionally nothing more than a commercial concern. It had a virtual monopoly on the exploitation of the country's economic resources, and in many regions acted as a *de facto* government. Where it was in its own interest, it suppressed or appropriated key local industries.

But local botanical knowledge was another matter. The company calculated that the Indian flora almost certainly contained plants of unrealized economic importance, and that local plant lore – and, for that matter, indigenous scientific knowledge among what was a far more educated workforce than in most of Britain's colonies – might be a short cut to discovering these. It began methodical survey work in the middle of the eighteenth century, and in 1787 set up a botanic garden in Calcutta.

Meanwhile, in the almost unexplored south-east coastal region known as Coromandel, a young Scots physician and botanist had already begun an impressive documentation of the region's useful plants. William Roxburgh had joined the company as a surgeon in 1776, aged twenty-five, but spent much of his time studying local plants. By 1789 he had abandoned medicine, and taken over the preparation of a book on the economic flora of Coromandel, started by a colleague. He had a good deal of material already: his own notes and a portfolio of illustrations by an Indian artist, whom he had kept 'constantly employed in drawing plants, which he accurately described, and added such remarks on their uses as he had learned from experience or collected from the natives'.

Roxburgh eventually accumulated more than 2,500 paintings for the company, and enlisted the help of Sir Joseph Banks, then director of Kew Gardens and hugely influential in eighteenth-century science, in getting them published. The twelve parts of *Plants of the Coast of Coromandel*, containing three hundred paintings, were published between 1795 and 1820, and remain one of the most extraordinary collections of flower paintings to have been published in Britain. The identity of the Indians who prepared them isn't known, except that they were mostly Hindus, and may have included artists such as

Haludan, Vishnu Prasad and Gurudayal, who are known to have worked for the East India Company.

But what makes the illustrations so exceptional is their quality as cultural hybrids. The company would have doubtless preferred accurate and unornamented field-guide illustrations, a crib for collectors and prospective growers. What they got was an exotic fusion of European precision and Mughal stylization that celebrated the formal structure and patterning of plants.

There was a long tradition of flower painting in Mughal culture, chiefly of delicate miniatures built up by layer upon layer of brilliant body colour. The paintings were finished by the use of very fine brushes, which were drawn across the paint to add texture or surface detail. In this way it was possible to suggest the lustre of petals or the leatheriness of leaves.

By modern standards Mughal flower paintings are exquisitely beautiful and highly successful at catching the jizz of a plant – not by impressionism but a kind of hyper-realism. But the East India Company regarded them, unmodified, as too obviously decorative. They lacked the literalism and austere clarity of line that had become customary in European plant illustration, and which were regarded as 'proper' for scientific representation. Nor did they employ techniques – use of perspective, for example – which in the West were regarded as essential for highlighting a plant's characteristic features.

So, as part of the process that led eventually to the chimerical style of *Coromandel*, company officials began training Indian artists in European techniques. One of the models offered was James Sowerby's work for the *Flora Londinensis*, with its fastidious recording of detail and attention to all the plants' internal and external structures. They introduced the artists to the subtleties of watercolour, and suggested how their own bright pigments might be muted for more restrained British eyes.

The Indian painters were happy to oblige. Working for the company was at least regular, even if they were paid only a pittance. But although they succeeded in achieving the kind of accuracy demanded by the company, the habits and traditions of their culture couldn't easily be suppressed, and the house style that evolved – which came to be called 'Company Art' – is uniquely cross-cultural. On the surface the paintings are neat, comprehensible, even diagrammatic where necessary. But working inside these conventions, and without ever falsifying the plant's detailed identity, the painters made *pictures*, relishing the contrast of colours and shapes, looking for suggestions of order, portraying an intriguing surface detail so sharply it looks as if it has been etched. The water chestnuts, *Trapa* spp., are a family of floating aquatic plants whose tuberous roots, rich in starch and fat, are a staple in Asian cooking and increasingly familiar in the West. The Hindu painter of *Trapa natans* var. *bispinosa* ignores the root, but is carried away by the floatiness of the leaves. Shaped like spades and edged with dark hatching, they are fanned out from the stem like a hand of cards. The Lagerstroemias, or crape myrtles, are now popular garden shrubs, but they're renewed in the *Coromandel* painting (of *L. reginale*), where the artist has emphasized the cut-and-crimped-paper appearance of the flowers, and elegantly arranged them as a chaplet round the shoots – though he's also rather awkwardly twisted one of the leaves round so that the underside is shown. A reluctance to be as literal-minded as their teachers hoped was quite common. Shading sometimes appears on the wrong side, and backgrounds can be so intense that they overpower the main subject. The illustration of Sappan (*Caesalpinia sappan*), which positions the small yellow flowers against the ferny leaves, has the look of a piece of exotically printed fabric, but hardly shows off the flowers to best advantage. Large leaves especially were apt to be painted in flat, unmediated greens, with an appealing eggshell finish but not much leafy realism. 'Most abominable leaves for which Master painter shall be duly cut', reads a severe company jotting on the back of one of the original pictures.

The lasting impression these paintings give, beyond the inventiveness of their patterning and layout, is their sense of light and space, which artists from the smaller and drabber landscapes of Europe could never hope to achieve.

Two women artists especially, working three centuries apart, exemplify the alternative tradition of plant exploration. They both began their travels comparatively late in life, and both were drawn towards Amazonia.

Maria Sibylla Merian was born to a respectable and religious family of Dutch artists in 1647. Fifty years later, separated from her husband, she was on a boat bound for Suriname with her daughter, with a burning ambition to paint the insects and plants of the tropical rainforest. Even if her paintings had been of no account, her adventurousness would have been. As it was, she introduced a new sense of the interconnectedness of nature to seventeenth-century painting.

Merian's early work was in the tradition of the decorative Florilegium, portraits of single flowers, bouquets and nosegays, painted attractively and accurately but without much originality. But in 1679, she published her *Raupen*, whose title in English was *The Wonderful Transformation and Singular Flower-Food of Caterpillars . . . Painted from Life and Engraved in Copper*. In each of the hundred plates one or more insects were depicted in their various life-stages: deep in a cocoon, feeding as a caterpillar, flying as a mature butterfly. And in every stage they were connected with a plant. She even showed the holes eaten out of the leaves by the foraging caterpillars, and made no attempt to fabricate some idealistic portrait of 'perfect' flowers. Her insects and plants were telling a story of interdependence, of process, of the value that was contained within their own circle of existence, without reference to the human world.

This was a radical break with the conventions of the time. Such insect books as there were (for example, Merian's neighbour and contemporary Jan Swammerdam's *General History of Insects*) dealt almost exclusively with anatomy and classification, not at all with the behaviour of the creatures, let alone their interactions with other species. Where insects and plants did appear together – for instance, in calendar paintings and Dutch still-lifes – it was invariably to convey some biblical or allegorical message.

There was an originality about the order in which Merian arranged her paintings, too. There is no obvious scientific scheme at work, and the paintings don't follow any known taxonomic order. Her biographer, Natalie Zemon Davis, suggests that 'Merian's goal was simply ill-served by boundary classifications. Her subject was a set of events . . . And to represent them meant crossing the line between orders and putting the plant and animal kingdoms in the same picture.' She chose instead an aesthetic narrative, guiding the reader through these 'transformations' using allusion and visual resonance. The end result gives a much more striking view of the connectivity of nature than if she had used conventional classification.

In 1685 Maria Merian's life went through a metamorphosis of its own. She fell out with her family, left her husband, and converted to a radical Christian sect, the Labadists. Her spell in the community lasted only five years. Disenchanted by the Labadists' monasticism, missing her work and artistic and scientific stimulation, she returned to Amsterdam and rarely mentioned the sect again. But as Natalie Zemon Davis points out, her spell at the community was 'a time of chrysalis, of hidden growth and learning for a woman who could not be pinned down'. She increasingly found that portraits of the plants and insects which she loved were themselves hermetic, wrenched out of their living contexts, devoid of growth and process. So in the 1690s she 'was moved to take a long and costly journey to Suriname' to paint at first hand the lives and transformations of the creatures whose inert forms were cropping up more and more in the collections of returning explorers.

When she set sail with her 21-year-old daughter Dorothea, without any protective men in attendance, and without any of the patronage that usually supported botanical explorers and artists, it was an expedition quite without precedent. Maria settled with Dorothea in a house in Paramaribo, in the small Dutch settlement, and immediately began work collecting, breeding and drawing insects. She found the native Amerindians and the African slaves much more helpful than the European planters. They took her exploring in the rainforest, explained the traditional uses of plants to her, and brought her maggots, chrysalises, and, once, a host of lantern flies, whose fiery lights and 'hurdy-gurdy' (*Lierman*) music enchanted her. Although she had a few slaves of her own, she was uncomfortable with the Africans' status, and especially with their treatment in the sugar plantations and mills. She constantly criticized the European obsession with sugar, pointing out how it had ruined the agriculture and social structure of the country. In Suriname, even taxes were assessed in terms of sugar: fifty pounds for every person over twelve, whether white, black or red; twenty-five pounds for every child.

Maria sketched from life, on the spot, and later worked up her drawing into paintings on vellum. Even then, she seemed close to the marrow of Amazonian life:

> When I painted, [wasps] flew before my eyes and hummed around my head. Near my paint-box they built a nest out of mud, which was as round as if it had been made on a potter's wheel; it stood on a small base over which they made a cover of mud to protect the interior from anything unsuitable. The wasps bored a small hole in it for them to crawl in and out. Every day I saw them carry in small caterpillars, doubtless as nourishment for themselves and for their young or worms [grubs], just as ants do.

But after two years, she could not bear the tropical heat any longer ('I almost had to pay for it with my life,' she wrote to a fellow naturalist), and returned to Europe, with a small collection of the butterflies, moths and larvae that she had not so far been able to paint. In 1705 her *Metamorphosis of the Insects of Suriname* was published in Amsterdam. Perhaps its most challenging single illustration, compared to the fashion in other contemporary natural history books, was 'Spiders and ants on a guava tree', a picture buzzing with tropical life. A bird-eating tarantula is mantling a hummingbird. The guava has ripe fruit but raddled leaves, the remains of which are being ferried about the twigs by processionary ants. One of the spiders is catching ants in her web, while another, smaller individual appears to be under attack by an ant. Life goes on, and goes round.

Yet, as in her earlier volume, *Raupen*, it is the ordering of her plates that perhaps has the most modern significance. Again, it is not scientific but aesthetic – or perhaps, more accurately, theatrical. She artfully moves the reader back and forth between the familiar and the strange. The opening plate is of a pineapple (already well known in Europe) and a vast cockroach, somehow suggesting the indiscriminate mixture of abundance and voraciousness in the wild tropics.

Like Maria Merian, Margaret Mee's upbringing inclined her towards an individual approach to painting. Born in 1909, in the depths of the Buckinghamshire countryside, she was educated at home by an aunt, a children's book illustrator with feminist leanings. Margaret had an obvious talent for drawing and went, for a while, to Watford School of Art. But she soon became involved in radical and trade union politics, and for the next twenty years painted only spasmodically. After the Second World War she enrolled in Camberwell School of Art, where she fell under the spell of Victor Pasmore. Pasmore was the founder of the Euston Road Group, who were in revolt against the fashion for abstraction in painting, and his own passion for naturalistic structure

– 'Look at the shapes – fit the shapes between the spaces' – was to have a profound influence on the sense of composition in Margaret's remarkable Amazonian landscapes.

She first travelled to Brazil in the late forties, to visit her sister, and settled in São Paulo with her husband in 1952. She made many trips into Amazonia during the next forty years, which are documented in her diaries. The most remarkable was in 1988, when she travelled up the River Negro in the hope of becoming the first artist to paint the night-flowering cactus, *Strophocactus wittii*, in the wild. The moonflower, though it had been grown in European conservatories since the nineteenth century, was the stuff of legend. Nothing at all was known about its life or natural history, except that it bloomed at night, with the release of an overpoweringly sultry and seductive smell. Only two paintings of the plant existed, both made 'in captivity'. Pierre-Joseph Redouté painted *The Queen of the Night* for Marie Antoinette, as it bloomed 'before the assembled Court'. Another portrait, entitled *The Night-Blowing Cereus*, appears in Robert Thornton's grandiose 1811 coffee-table book, *The Temple of Flora*. Its portrayal, stage-directed by Thornton, shows something of the way in which exotic flowers were perceived at the time. The cactus flower itself, looking as if it has been dipped in honey, is the work of the journeyman landscape painter Jonathan Reinagle. The background, which includes a dark English wood, a gothic tower with the clock set at midnight and the moon peeping from behind a tree, is attributed to another small-time painter, Abraham Pether.

Margaret Mee and her party were only a couple of days upriver from Manaus when she saw the first *Strophocactus*, its 'scarlet leaves . . . pressed against the trunk like transfers'. More followed, including tresses of a plant blown close to the river by a storm. They found another in ripe bud on a tree covered with epiphytes, and settled down for their vigil just as the sun was setting. Margaret's attention to a plant whose life is governed by the sparse light of the moon was fastidious: 'Within an hour the sky was black and only the brightest stars showed through a layer of thin cloud covering the silent forest. We took turns to keep watch on the buds using the torch beam briefly, so that we would not disturb the opening. Once in the confusion of the dark the torch slipped over the side of the boat and we watched its powerful beam slowly dim as it fell through the deep, tea-coloured water.'

Two hours later, the first petal began to move, then another, until the whole flower was rippling into life. Margaret moved a chair up on to the roof of the boat, and her companion Sue held a small battery light over her while she sketched. But there was a film cameraman with her also, and she worried that his strong photographic lights were slowing the blooming. The lights were dimmed, and she continued to work by moonlight. The bloom was fully open in less than an hour, and as Margaret continued to sketch, she hoped the pollinator of the cactus – either a bat or a large hawk-moth – might arrive. But nothing appeared except clouds of flies attracted by the heady perfume, and Margaret fretted that 'our intrusion had deterred the pollinator, upsetting the delicate balance between plant and the animal which has taken tens of millions of years to evolve'. She imagines the effects of such intrusions multiplied all across Amazonia, and finds them 'too much to contemplate'.

It is not clear from Margaret Mee's text whether she was aware then of the full reproductive cycle of the moonflower. It is an extraordinary story. Once pollinated, the large red fruits fall into the river floodwaters where they are eaten by fish. Once the seeds have passed through the fish, they float away, buoyed up by internal air sacs. The seed lodges against a tree trunk and germinates, and begins its climb up to the canopy again. A canopy species dispersed by a fish is a bizarre and incongruous image, but emblematic of Amazonia's knitting of wood and water. And Mee's finished portrait of *Strophocactus* seems to intuit this story. The painting is a masterpiece, and, I suspect, makes a small tongue-in-cheek dig at *The Temple of Flora*'s melodrama in its positioning of the moon in the top right-hand corner. But Mee's is no shy English moon, sheltering behind some baleful elm. Instead she fills the top of the painting with a vast amorphous disc of white, that bathes her background trunks with submarine light. In the foreground, four moonflowers, sharp-petalled and pure white – their proper colour – wave like sea-urchins. The rust-coloured foliage is clamped against the trunk of its host tree, edging upwards. This is a portrait not just of the night but of Amazonia itself, of a universe rising out of the waters.

Crown shyness

Botanists – ESPECIALLY AGEING BOTANISTS – are ground-peerers. The variety and diminutive tangles of herbaceous flowers (combined with the inexorable advance of that crick in the neck) have always directed their gaze downwards. From the seventeenth century onwards, illustrations and cartoons of plant-lovers have them on their hands and knees, in supplication to the tufts of the earth. It is no coincidence that the localities of some of Britain's scarcest trees – large-leaved lime, black poplar, wild service tree, not exactly shrinking violets – are still being uncovered in the twenty-first century. No one thought to look up.

At Eden, the domes themselves draw your eyes up. Yet it's still hard to keep your attention fixed in the canopy. Not expecting anything to be there, I almost miss two colobus monkeys scrambling through the foliage about thirty feet up. They're gone in a trice, but about an hour later I meet them again, loafing about the broad trackway near the entrance to the biome. They're performance artists, of course, not real monkeys: Eden's hostility towards becoming a zoo isn't going to vanish just like that. But the detail of their costumes and their body language, and their piercing, primate eye contact, give them a compelling presence. These actors have watched colobus at close quarters. They squat on their haunches, arms looking impossibly long for humans, and gaze about in quick, jerky glances. One gets up on all fours, and, every limb stretched out straight, advances towards a small boy. He's about five years old and holds his ground, giggling uncontrollably in a mixture of excitement and uncertainty. He knows it's not a real colobus. But it's not *un*real either, not a computer graphic or a creature from a wildlife documentary. It may be a human, but that doesn't stop it being a *live animal*, unpredictable and mischievous.

By now both monkeys are working the crowd, untying shoe laces, emptying kids' bags on the ground, rushing off with anoraks. The children are in paroxysms of laughter and discomfiture. The youngest are stamping their legs up and down, not quite knowing how to respond. The braver ones stretch their hands out, and touch the monkeys' fake fur. Their grown-up composure, pumped up for this day out, is in tatters. These artfully intimate performers have taken them to limits that no longer occur in their lives. The monkeys are the strangers (and the strangeness) by whom children are no longer allowed to be touched. They are symbols of the wild in which they're no longer allowed to play. If the children have been models of restraint and orderly behaviour in their walks round the biome, the tropics have now reached out and snaffled them, in the most benign of lianas.

Swamp Circus, a collective of dancers and acrobats with a commitment to ecological themes, are down at Eden for Canopy Week. It is a felicitous pairing. Eden has a four-year rolling programme to celebrate the rainforest canopy, the least explored, most complex and most mysterious ecosystem on the planet, and so far removed from our earthbound reflexes that it needs a sense of theatre and the absurd to open it up. Later I watch another pair of Swampies (as they style themselves, in barely disguised tribute to that great eco-warrior) transformed into plumed basilisk lizards. This is a different kind of performance – stately, controlled, alien. The lizards materialize on the ground under the trees, scarcely noticed until they begin to move. They lift their feet meticulously, their lizard masks tilted up towards the audience while their human gaze is fixed on the ground. They climb ropes up into the canopy, and hold astonishing lizard-like poses. The audience's mouths are open, not just at the acrobats' athleticism but, one feels, at the lizards' too, at the dignity of lives lived almost invisibly in the treetops.

The rainforest canopy was not only unexplored until a few decades ago, but barely acknowledged as an ecosystem in its own right. Access to it had traditionally been achieved in the manner of early orchid hunters and modern loggers, by bringing it summarily down to earth – at which point, of course, the system disintegrates and ceases to exist. It thrives by intercepting over 90 per cent of the sunlight that falls on the forest, and is believed to contain between 40 and 95 per cent of all the plants and animal species on earth.

The breakthrough came in the early 1980s, when a team of scientists from the University of Montpellier in France had the idea of approaching the canopy from above. They would settle gently on the surface in the basket of a hot-air balloon, and with luck cause the minimum of disturbance. Soon they had developed a raft which could be lowered from a balloon, or a more manoeuvrable airship, giving its human occupants the chance to descend on to the canopy themselves. It was made from inflatable rubber tubing and nylon netting, in the shape of a starfish, or the end of a ski-stick. The rubber 'sausages' were wide enough to walk along, and up to five humans at a time could scramble about the PVC netting.

The complexity of life that was found through this 'treading lightly' approach was a revelation. The very biochemistry of the plants was more complicated than that found at ground level, probably because the insect life is so much richer, and has generated, by co-evolution, a comparable variety of chemicals in the plants, to attract and deter it. The flowers of *Napoleonaea vogelii*, for instance, have a complicated structure and chemistry, and different parts of them are attractive to different species of insect. The pollinators are tiny insects called thrips, but the flower attracts other insects that have no obvious connection with pollination, for instance moths of the *Glyphipterix* family. These look and move like a family of ferocious jumping spiders called salticids, and may consequently deter other predatory insects. Such symbioses are common high above ground level, because resources can be too scarce and specialized to be fought over. This is the domain of the epiphytes, plants which aren't rooted in the soil but attached to the limbs of trees (or to each other), and live off sunlight, the moisture in the air and the nutrients brought down by rain-trickles, or in bird-droppings. Orchids are the best known (and make up two-thirds of all known epiphytic species) but there are lichens, mosses, cacti, bromeliads. Tillandsias have stringy leaves covered with small scales which draw in water from the air. Some bromeliads are 'tank-plants', equipped with watertight 'buckets' which can contain several litres of water. These not only act as reservoirs for the bromeliad itself but as small-scale freshwater environments for other plants and insects. Sometimes there are many layers of mutual dependence. Carnivorous plants like bladderworts sometimes live inside the buckets, for instance, surviving on minute aquatic animals.

The weight of epiphytes sometimes exceeds that of the tree to which they're attached. But the trees get something out of the relationship, too. They are protected against dehydration by the swaddling of mosses and creepers. Each tree sustains a huge population of mutually balanced animals, which may feed on the epiphytes and pollinate it. And recently canopy researchers have found that under the thick mats of lichens and mosses, the host trees send out short aerial roots – maybe a hundred feet above the ground – which capitalize on the pockets of decaying leaves and debris trapped around them. Epiphytes don't need soil, but succeed in making minute quantities of it.

Many of these new discoveries show how inadequate ground-based and agriculturally tinged theories are to explain a system surviving with minimal nutrients way above the earth. And one discovery in particular challenges many ideas about the way in which organisms in close proximity are supposed to compete. At the very top of the canopy the crowns of neighbouring trees seem to gather themselves in, and leave a visible gap of perhaps half a metre between them. Buds stop developing around the edges of this zone, well before the trees actually touch. No one yet knows the mechanism which sparks this show of mutual deference. It may be some subtle alteration in the amount or quality of light received by the buds as they approach the neighbouring tree, or some airborne growth-retarding chemical, similar to those given off by many plant roots. The scientists who discovered the phenomenon have given it the rather touching name of 'crown shyness'. Yet it seems analogous less to shyness than to politeness, an ecological etiquette that helps ensure the diversity from which all species benefit.

Recently, a Dutch company called Kronendak have begun to explore the concept of canopy farming. The ethical motivation is to try to halt the continued destruction of rainforests by increasing the value of their non-timber products. This is not about a return to hunter-gathering. Medicinal, food and ornamental species would be deliberately cultivated *in situ*, hopefully providing employment opportunities for local communities as well. The commercial motivation is of course the escalating value of scarce tropical plants as they vanish along with their habitats. Kronendak argue that 'such efforts would contribute significantly to the conservation of the canopy habitat, precisely because they need the sustenance of the living and intact rain forest in general', and list their objectives as follows:

> To establish and manage sustainable rain forest ecosystems which –
> Provide beneficial returns to the local populations;
> Create vibrant conservation areas;
> Create wealth and improve the quality of life;
> Establish, demonstrate and expand the standards for sustainable silvicultural technologies including canopy farming.

A managed rainforest is better than no rainforest at all. But it's worth looking at the compatibility of these objectives before seeing canopy farming as a panacea, and remembering that (as we've already seen) most ecologically sensitive farming systems are eventually compromised by purely commercial decisions. Would, for instance, the cultivated species be impressed into the canopy at the expense of the species growing there anyway? What if they are eaten or damaged by indigenous insects or birds? Will there be 'pest' control? What will be the likely disruption to the whole ecosystem of the cultivation and harvesting activity? Canopy farming could be a boon, but it could also be a severe test of our ability voluntarily to control the intensity of horticulture, and of the maturity of our ethical stance towards organisms not obviously of use to us.

THE
THIRD EDEN

Cultural archaeology

A^{T} THE EXTREME SOUTH-EAST OF EDEN, there are a few remnants of what the Bodelva quarry was like before its restoration. The bare granite walls, draped with curtains of gorse and oak, fall to a ground-base that, before the Project began, was deep under water. Now it's been drained, and covered with grass. But every time the excavators dig into this floor to add another extension to the great botanical theatre, they expose these layers of Eden's past – and also the futures for this space which did *not* come to pass. History is about untaken options, roads not travelled, as well as about achievement and realization.

The huge rib of granite that underlies Cornwall from Dartmoor to Land's End was forged 500 million years ago, from lava pouring out of a string of undersea volcanoes that stretched as far north as Scotland. No one is sure whether granite is formed directly from the setting of molten lava, or whether it's created from existing rocks which have come into contact with hot magma, causing them to melt and then recrystallize. Either way, the rock forms and cools deep beneath the earth, and is only seen on the surface (as in the Dartmoor tors) where erosion has stripped away overlying rocks. Its slow, subterranean origins are evident in the large size of the resultant crystals – the semi-translucent white or pink grains of quartz and felspar, and the sparkling silver flakes of mica.

But in south-west Cornwall, and at the site that would eventually become Eden, something more elaborate happened to the crystallizing granite. In places, especially where there were cracks in the solidifying rock, the mica and felspar were attacked by the superheated steam and carbonic acid given off when molten magma comes into contact with water. This converted them into the powdery form of kaolin known as china clay, which was deposited in almost circular 'pipes' along the lines of the fissures. There it lay till less than three hundred years ago.

Meanwhile, Cornwall went through the turbulence of aeons of geological history, spending much of the time under either the sea or swampland, and being variously elevated, dried out, warped and, with the rest of Britain, shunted northwards from its original site just two hundred miles north of the equator. By the end of the last glaciation, fifteen thousand years ago, its boundaries and contours were pretty much as they are today.

Five thousand years on and what was to become Bodelva Moor was covered with forest, a woodland not unlike the ancient woods that still survive in the West Country, though probably rather drier, with sessile oak, hazel, rowan and perhaps wild service tree. That might have been one viable future for the spot, to continue as woodland, as some areas of Dartmoor did. But by 1000 BCE, much of the woodland was gone, cleared by Bronze Age farmers. The climate was becoming rainier and cooler at the same time, which inhibited the regeneration of trees and encouraged moorland to develop. It was probably about this time that small-scale open-cast mining began for tin, lead and coal. But it was not until the medieval period that this was instituted on an industrial scale, and not until the early eighteenth century that the china-clay beds were first exploited. The opening of vast holes on Cornwall's surface had begun.

By this time the intellectual tradition that would make places like Eden possible had also taken shape, and owed its own mite to the myth of the Lost Paradise. The realities of the New World had finally put paid to the notion that the original Garden might still survive intact, and the repeated discoveries of prodigious plants and creatures in the four corners of the globe were challenging the assumptions about creation presented in scripture and the classics. Nature, far from seeming irredeemably corrupted, looked rather well, and full of limitless possibility. The idea grew that, at the Fall, nature had not so much been cursed and destroyed as scattered across the earth – to be discovered by mankind when it was ready. It therefore became one of humankind's great redemptive tasks to reassemble the jigsaw puzzle, and make coherent sense of it. Just as discovering the Garden of Eden was a covert motive behind the economic projects of the early colonists, so rebuilding it was one of the aims behind the development of botanic gardens in the sixteenth and seventeenth centuries. Their full purposes, of course, were complex and often local. The first, founded at Padua in 1545, had links with the mercantile interests of nearby Venice. The Chelsea Physic Garden in London was notionally devoted to exploring and teaching plant medicine. The Jardin du Roi in Paris (1626) was a huge celebratory collection, which came, Sir Thomas Baskerville reported, from 'the remote Quarters of the World', and contained plants 'comprehended as in an Epitome'. But behind all of them was a belief, in tune with the optimistic politics of the time, that a model of Paradise could be rebuilt. Man's toil, solemnly imposed at the Fall, was redefined as a virtuous pleasure, if it could be carried out in the elevating setting of a garden. Peter Smith, a member of Samuel Hartlib's circle of idealistic reformers, wrote in 1657:

> ... which indeed, though it was a punishment to Adam, yet it is
> the best nurse of health, & chearfullnessse to his posterity: And
> I suppose that if great conquerours, & troublers of the world were
> but sensible of those pure & naturall delights of eating the fruits
> of their owne planting & living upon their owne labours, or in a
> higher sphear were capable of the joy that ariseth from such
> actions that benefitt the publick without hurt to any particular,
> such as are the invention of the plow, planting of fruit, dreining of
> Fens, etc. they would acknowledge their glory to be very partiall,
> & their great joyes to be attended with intervalls of trouble &
> remorse in comparison of these.

This rather touching combination of down-to-earth common sense and an unqualified belief in the benefits of technology catches the mood of the early period of the Enlightenment. John Evelyn echoed it when he wrote in his unpublished manuscript *Elysium Britannicum* that 'Adam instructed his Posteritie how to handle the Spade so dextrously, that, in processe of tyme, men began, with the indulgence of heaven, to recover that by Arte and Industrie, which was before produced to them spontaneously; and to improve the Fruites of the Earth, to gratifie as well their pleasures and contemplations, as their necessities and daily foode.' Indeed, Evelyn himself – royalist, gardener, landowner, jobbing writer – could well stand as an emblem of the heady optimism and political (and ecological) naivety of the late seventeenth and eighteenth centuries. Like many of his contemporaries, he believed that new techniques of gardening might restore the fertility of the earth that had been lost at the Fall. The proposal for the *Elysium* begins with an account of Adam's exile, and ends with the Kings and

Philosophers who, 'when they would frame a type of Heaven ...
describe a Garden and call it ELYSIUM'. Evelyn made no bones about
the beneficiaries of this New Eden; they were 'Princes, noble-men and
greate persons who have the best opportunities and effects to make
gardens of pleasure'. His *Sylva*, a work of some expertise on forestry,
was purportedly written in response to a fictitious timber famine
(see pp. 64–5) but in reality served as a blueprint for the eighteenth-
century incursions of commercial forestry into park and commonland.
In *Pomona* he drew up a plan to plant a fruit tree every hundred
feet throughout the country, and proposed to surround London
with regular enclosures of thirty to forty acres in extent filled with
scented flowers. The one portion of his *Elysium* that was published – a
precocious and waspish tract entitled *Acetaria. A Discourse of Sallets* –
coloured in his conviction (widely shared at the time) that fruit and
herbs, being Adam's diet in Eden before the Fall, were the foods
that humans bent on redemption needed to adopt. 'Certain it is,
Almighty God ordaining Herbs and Fruit for the Food of Men, speaks
not a Word concerning Flesh for two thousand Years ... And what if
it was held undecent and unbecoming the Excellency of Man's Nature,
before Sin entered, and grew enormously wicked, that any Creature
should be put to Death and Pain for him who had such infinite store
of the most delicious and nourishing Fruit to delight, and the Tree of
Life to sustain him?' Evelyn was, by all accounts, no vegetarian himself,
but he was ferocious in his attacks upon meat-eating and cruelty to
animals. A diet of fruit and herbs, he believed, was not only more
natural but was conducive to an Edenic life of longevity, health,
frugality, leisure and ease: 'the Hortulan Provision of the Golden Age
fitted all Places, Times and Persons; and when Man is restor'd to that

State again, it will be as it was in the Beginning'. And the path to this restoration was not through revelation or supplication, but through human enterprise and the philosophy of Improvement. 'It has often been objected, that Fruit, and Plants, and all other things, may since the Beginning, and as the World grows older, have universally become effete, impair'd and divested of those Nutritious and transcendent Vertues they were at first endow'd withal: But [there is not] the least Decay in Nature, where equal Industry and Skill's apply'd'. As Timothy Nourse wrote the year after the publication of *Acetaria*, gardening might 'most properly be call'd a Recreation ... from the Restoration of Nature'.

Evelyn and his contemporaries had, in effect, written the first outline script for the Green Revolution, a manifesto on the salvation of humans through the improvement of plants. But, like their successors, what they hadn't thought through was which of the planet's inhabitants were to be the inheritors of this new paradise.

In the eighteenth and nineteenth centuries the drive towards Improvement became more secular and commercial. The garden became the ostentatious landscape park and then the industrial farm, and Evelyn's rather tender concern for domestic animals was forgotten. Redemption through profitable agriculture metamorphosed easily into salvation through trade and industrial progress, nowhere more so than in Cornwall. At the very end of the nineteenth century, the Bronze Age field systems around Bodelva were dug into for their china clay, to supply the burgeoning porcelain factories in the Midlands. That was always going to be a developmental cul-de-sac, because the raw material was finite.

Meanwhile, spurred on by a cooling of the climate during the nineteenth century, the Wardian case was joined by the full-blown conservatory, an oasis of warmth in which yearnings for a tropical paradise could thrive. Already a pattern was developing, in which the outside, 'natural' world would be the site of exploitation and industry and improvement, and 'nature' itself (and dreams of more ideal worlds) would be realized in controlled enclosures. When Tim Smit had his dream of the world's biggest conservatory, it was absolutely in that eighteenth-century tradition of the re-creation and 'Restoration of Nature', to try to partially rejoin these two traditions, and to site a theatre of plants in the scars of a worked-out industrial quarry. The ambitions for the place took shape before its site was discovered, and began with a list of all the plants 'that mankind had used and deserved a place'. They would be grouped together according to their rough vegetation types and climate zones – their 'biomes'. There would be a 'Roofless Biome', containing a range of plants from the world's temperate zones, capable of growing outside in Cornwall; a Humid Tropics Biome, with plants drawn from Amazonia, West Africa, Malaysia and the Oceanic Islands; and a Warm Temperate Biome, filled with the vegetation of the Mediterranean and similar zones across the world, such as California and South Africa. The exhibits for each biome would set out to demonstrate how we have gradually farmed and domesticated the earth's plants, but also how over-intensive farming, development, pollution, tourism and ever-increasing demand ('whether it involves razing rainforests or wanting mange-touts in winter') were straining that ancient partnership between people and plants. One way and another, Eden's agenda already had humans centre-stage.

When it came to finding a site for the Project, there was never much doubt among the team assembled by Tim Smit that it had to be somewhere in Cornwall. The combination of a benign, humid climate in which a huge range of plants could prosper and a landscape (and economy) in urgent need of regeneration was a coincidence that couldn't be ignored. Eden, with the help of 'Arte and Industrie', could be a catalyst for new opportunities in local horticulture and environmentally based business. Bodelva china-clay pit – a crater sixty metres deep and spreading across an area equivalent to 350 football pitches – was a gauntlet that fitted the team's ambitions. 'An architect would fall over backwards wanting to build something in it', remarked one of the planners. Everything about it was a challenge – the unstable edges, the flooding of some of the deeper parts (natural springs bubbled out of the slopes), and 'last, but hardly least, it contained no soil, only sterile muck – a problematic element when the ultimate client is a plant.'

I have to intervene with a footnote here, because much of what Eden does – and what humankind does on the world's larger stages – is predicated on this belief in the importance of soil fertility, a myth that, as we've seen, goes back to the earliest days of agriculture. Yet unless it is actively and continuously poisoned, there is no such thing as 'sterile muck'. All bare substrates will eventually be spontaneously colonized by plant life, as they were at the very beginnings of terrestrial life, and have been many times since. The digging of a china-clay pit is small stuff, after all, beside the scourings of glaciers. The process can be witnessed in many of the unrestored pits in the area – the first smudges of algae and mosses, springing from wind-blown spores and nourished by the carbon and nitrogen compounds in the rain, then the first small annuals rooted in the edges of the moss clumps. Meanwhile, round the lip of the bare quarry, the root systems of the surrounding vegetation creep in – grasses, bushes, then full-sized trees, whose fallen leaves begin to fill in the pools and flashes, helped by landslips from the unstable quarry sides. It is, of course, a very long-drawn-out process: the natural filling of the fifteen-metre-deep lagoons at Bodelva might have taken tens of thousands of years. That is a timescale in which whole civilizations can rise and fall, and makes waiting an impractical business for us. But it's crucial to understand that the process of natural colonization is inevitable and inexorable, and that the problems are to do with our life-rhythms, not nature's regenerative powers.

So, a helping – an *accelerating* – hand was given. The one and a half million tonnes of spoil heaped around the site were shovelled back in, and much of the site levelled off. The sheerest, most unstable faces of the quarry were pinned in place by 2,000 twelve-metre-long rock-bolts. Eighty-five thousand tonnes of soil, mixed together in bespoke formulae from compost and clay in another abandoned pit nearby, were strewn across its now more regular surfaces.

The design of the greenhouses emerged from big-scale thinking, too, much influenced by science fiction. One reference that kept surfacing was the cult 1970s film *Silent Running*, set in a future in which all the earth's natural vegetation and animal life have been destroyed in a nuclear holocaust. In the film, a cluster of plant-filled spacecraft – part arks, part biomes – orbits Saturn, awaiting a time when Earth may again be able to support life.

In the end it was perhaps inevitable that the successful design was inspired by organic forms. The geodesic dome is based on interlocking

hexagons – an archetype that occurs throughout nature, most notably, as we've seen, in the honey bee's comb. This isn't an accident. The uniform distribution of forces and stresses in a group of hexagons makes it the strongest and most economical way of arranging layers of matter, especially in round structures. And, as a form in which hard science meets not only natural design but traditions of spiritual belief in the perfection of round bodies and the significance of six-sided things, geodesic domes became hugely popular in the alternative cultures of the 1960s. Circuses and environmental seminars were held in them. In the south-western deserts of the US in the early '70s, they were built on the back of pick-up trucks and the tops of cars. Mobile summerhouses.

Eden's plants, too, came initially from the same kind of do-it-yourself culture. They were found at car-boot sales and garden centre clear-outs. They were donated by local people who wanted a good home for unloved house plants, and by prestigious botanic gardens.

Now most are raised from seed or cuttings, in a nursery at Watering Lane (five miles from Bodelva), where the Eden Project first set up shop.

In many ways the entire Project is based on this kind of marriage between high technology and counter-culture make-do, between the intensity of a think-tank and the laid-back ambience of a rock festival. Which is why, no doubt, it is so agreeable just to *be* there: it has matured beyond being just a 'site' into being a real place. A friend who wrote to me after her first visit said simply, 'I want to go and live there.' I know what she meant. During my own longer visits, bridging all four seasons, I began to feel the kind of fond and deeply irrational attachments to spots and moments that one develops as a child for special corners of the house and ritual routes to school. In the early evenings, I'd always drift to the centre of the Mediterranean display, sit down on one of Dominic Cole's Moorish walls and watch the thinning crowds go by. When they're emptied of people the biomes are transformed into spaces full of light and of teasing, fugitive sounds you haven't noticed before. I had favourite outside strolls, too. One, down the slope from Wild Cornwall towards the biomes and restaurants, had a palpable feel of going home after a day out.

I enjoyed perching in the eating areas. They seem to create opportunities for visitors to begin to talk beyond their tight family groups. Interaction between customers is limited here, and maybe a consequence of Eden's irreproachable standards of safety and smooth-running efficiency. Its visitors may read about our common human problems on every poster-bearing space, but they rarely *experience* any, even of the most domestic kind. No groups of visitors have to come together to negotiate a flooded path, or to try to disentangle the workings of a wilfully unlabelled display. Eden's internal perfection, the consummate satisfaction it gives to visitors, paradoxically belies the disorder of the world outside, whose flaws and challenges it is Eden's mission to explore. Perhaps the Project does indeed see itself not just as a route to a solution, but as part of the solution. Could we all come and live here? Is the ultimate future of humankind's partnership with plants to be found in some immense extension of the Eden model, in a benign, enclosed paradise, a Wardian case that includes humans too? Or, less fantastically, in a metaphorical enlarging of the biomes, so that all the planet's vegetation is responsibly and sustainably gardened, brought into our domain? If so, what is to become of the wild, from which the cornucopia of humanly useful plants emerged? What will be its place, its purpose, its *point*? At precisely the moment when we are finally grasping its astonishing complexity and independence, is the rest of creation to be evaluated purely in human terms again, as it was in the days of Genesis?

Those who plan Eden's progress sensibly avoid such philosophical wrangling. Pragmatically, they'd like the place to satisfy all the roles that are glimpsed in it, to be a shrine, an example, a provocation, a resort. But mostly they prefer not to see it as some definable end-product at all but as a *process*, with aspirations to be, like the organic world it celebrates, in a state of continuous evolution and interaction.

The Project prides itself, for example, on the greenness of its 'house rules'. The glass bottles that the visitors diligently poke in the recycling tubs are crushed to produce aggregates for road sub-surfaces. The Project's electricity comes from the Green tariff, which in Cornwall means (notionally, anyway) wind turbines. It's self-sufficient for most of its water. Yet it is a long way from being some kind of ecological utopia. It has no plans, as yet, to grow the food for its restaurants on an on-site organic farm, or to generate its own power. It doesn't want to be self-sufficient so much as to reach out. If it did come to a choice, it would rather be judged by its influence beyond the perimeter of the Project than by its own internal purity. It talks a lot about being a cat-alyst, an inspiration, a source of energy which may help people 'reconnect with nature'.

The team is fond of telling the story of the group of deprived inner-city youngsters who came to stay at Eden, and began talking about '*our* Project'. They were inspired less by the displays and the conservation messages than by the story of how Eden 'happened', how it created itself by an act of will out of apparent dereliction. Business leaders are also intrigued by Eden's can-do philosophy, and come to examine its very non-hierarchical structure, which the staff are apt to describe as 'fluffy' and 'soft'. New initiatives can spring out of any part of Eden's organization (except, so far, its visitors) and are as welcome as roses in winter.

Yet, however much it may shy away from the idea that it should present finished models, solutions, Eden cannot escape from the myth it has created around itself. Every signal – from the protective round-ness of the domes to the repeated messages about 'the partnership between people and plants' and the concentration on species that human beings have 'used' – reinforces the image of the world as humanity's garden, contained and nurtured for our benefit. The sense of an enclosing fence is every bit as strong as that of a paradise within.

Ancient myths have stressed this idea repeatedly. But never with total approval. Even the original Garden of Eden story is poignant about the loss of innocence and benign wildness at the Fall. A concern about what happens *beyond* the fence has hovered persistently, and has a new urgency today. Have we considered what the importance of unmanaged, non-useful plants might be? Do we recognize that the newly developed strains of, say, rice that can be grown at high altitudes do so at the expense of the indigenous plants of the hills? Does the breathtaking intricacy and mutuality of life being uncovered in the rainforest canopy make us ponder whether we have evaluated the 'usefulness' of plants correctly? All manner of species can be domesticated into food and fibre and medicine, but what about those that keep us breathing in the first place? It is easy to celebrate hemp and moonflowers, but can we find the same reverence for the oceanic plankton that are such major absorbers of atmospheric carbon dioxide and regulators of the climate? For the unappealing underground fungi whose mycorhiza generously help most of the planet's larger plants to grow? For the millions of modest insect species whose existence is still not even recognized, but which protect the flowers that other insects pollinate to produce the seeds which grow into the trees that supply the world's oxygen . . .?

And if we can, are we able to go one step further and see their intrinsic worth, as members of the planetary community, regardless of what they can contribute to our lives? We may talk of 'the partnership between plants and people', but the plants were not, so to speak, willing volunteers in that relationship. What might be their ambitions for their relationship with us?

What they face, along with us, is a breakneck process of change across the planet driven by population growth, business ambition and the desire for social justice. The world's population is thought (maybe wishfully) likely to stabilize at about eight billion, but what is to be done about social justice? At present, those living in the developed world have almost forty times the environmental impact per capita, in terms of resources consumed and pollution emitted, of those in the underdeveloped world, a gap which is chiefly the consequence of differences in standards of living. The drive to narrow the gap in those standards is relentless, fuelled not just by a sense of fairness but by capitalism's inherent opportunism. But how is it practically to be achieved? It is barely conceivable that the developed world would agree to lower its standard of living (especially as this is continually rising). Yet if the rest of the world's is to rise instead, and in consequence raise its environmental impact up to the level of the rich nations, the planet's life-support systems will not be able to cope.

The solution mooted for this conundrum is something called 'sustainable development', a process which is meant to deliver economic growth and prosperity at no cost to the planet's ecosystems. The 1992 Rio Conference's Agenda 21 defined it as 'meeting the needs of the present without compromising the ability of future generations to meet their own needs [meaning future generations of humans presumably, not all organisms]'. Tim Smit interpreted this as an agenda for Eden which involved 'exploring development in the fullest sense of the word: the sustainable development of human potential and the achievement of the optimum quality of life for all, across economic, social and cultural [but again, not species] boundaries'.

These are noble aims, but what sustainable development means a decade after Rio, and whether, indeed, it has ever been more than a theoretical dream, is less certain. It is hardly a programme any more, and scarcely qualifies as a myth. Its connections with magic are more superficial. It is an incantation, like Abracadabra! or Hocus-pocus! – a form of words designed to raise expectations rather than describe the world. Another piece of theatre.

Things looked compromised for the sustainable development idea from the outset. It grew from the more comprehensible notion of sustainable use, first applied in forestry: you took out of a forest no more woodland products than it could recoup in the same time. It was a responsible and uncontentious policy when it considered timber products alone. It was used, for instance, as a description of systems where harvested natural ('first-growth' or 'old-growth') trees were replaced, tree for tree, by planted saplings. Sustainability here referred solely to the timber resources, and took no account of the complex networks of fungi, insects, lichens and herbaceous plants that depend on old-established forest, and are permanently destroyed when this is replaced by plantations harvested as a long-term arable crop.

The same slipperiness and selectivity are evident in the way the phrase is used today. The UK government's house-building programme (subtitled 'Sustainable communities') rests on a view of sustainable development which has nothing whatever to do with environmental impact, the use of renewable resources and the like, but with the sustaining *of* communities, by, for example, the adequate provision of schools and policing, and policies designed to keep traffic flowing. Development agencies in poor areas of the world regard the provision of clean water and hydroelectric power as part of 'sustainability' regardless of the impact of water extraction on local water tables and ecosystems. Nature conservation bodies in the rich areas regard sustainable development as the provision of small, multi-purpose green spaces around a new motorway or airport, regardless of the unsustainability of the development itself. All commendable aims in themselves, but stretching the meaning of sustainability beyond common sense.

How is it that a seemingly unambiguous English word became so plastic in meaning? When words are abused like this, it's often illuminating to explore their roots. The earliest, sixteenth-century meaning of sustainable was 'capable of being borne or endured', an inflection which saw things from the point of view of the suffering person (or, we might say today, suffering organism or ecosystem). But at the time of the industrial and agricultural revolutions of the nineteenth century it came to mean 'capable of being upheld or defended', subtly switching the focus of sympathy to the oppressor or entrepreneur – or developer. These days sustainable seems to mean what you can get away with for the next twenty years.

'Sustainable' is not a word capable of qualification. You cannot have 'partially sustainable development'. The development can either continue indefinitely and benignly or it can't. In its original sense, of 'capable of being borne', sustainable development may be a contradiction in terms, except in the natural world, where it is the way in which all ecosystems evolve. In nature, 'development' proceeds according to three inflexible rules. All energy inputs come either directly from the sun, or indirectly, through sun- (and moon-) driven weather changes, such as tides and winds. All waste is eventually returned to and recycled inside the system. The input and output of both energy and resources are equal. Yet all ecosystems have an innate tendency to become more complex and diverse, even when their energy inputs and the land area they occupy are constant. The most elegant expression of sustainable development in nature is the way that bare land (including china-clay pits) progresses spontaneously to the condition of complex, three-dimensional forest.

It's possible to see how this process could be echoed in human development, and some experiments in forest farming and permaculture come close. But there is little sign of it elsewhere even in current best practice. Almost all human building materials – stone, sand, metal, any synthetics made from fossil fuels, any materials in fact except wood and other plant products – are non-renewable, and their use is therefore unsustainable. They will, sooner or later, run out. All energy sources – including 'renewables' like wind and solar power – rely on the same non-renewable materials for their hardware. Most overlooked of all is the fate of the ultimate non-renewable resource: the land itself, and especially the photosynthesizing, greenhouse-gas-absorbing, wild plantscapes of the planet, vanishing daily under cultivated plants as much as under new roads. These are strictly finite. Anywhere on the earth's surface it is possible to draw a graph of the disappearance of undeveloped land against time, and it crosses the zero axis in a finite number of years. This is a loss which unquestionably compromises the ability of 'future generations to meet their own needs' – let alone the ability of other organisms to meet theirs. It is hard to think of an agricultural system which even approaches sustainability. Organic farming, relying usually on petrol-driven machines for cultivation and product delivery, and occupying – usually irreversibly – land which once sustained wild ecosystems, certainly doesn't. Only as farming systems begin to resemble wild habitats – as in the cork-oak *dehesas* of southern Spain, and, perhaps, in future canopy farming – do they come close to sustainability.

Interpreting the idea of sustainable development this literally may be too pedantic and severe. *Any* moves, however compromised, towards meeting human needs while 'treading more lightly on the Earth' are to be welcomed. We are, after all, creatures of the planet, too, and have every right to our slice – even though, sooner or later, it will have to be a much thinner slice. And it may be that, despite the long-term threat it would pose to our own survival, we will choose to put our own immediate needs first. This is a moral choice we could make, honestly and openly – perhaps in the spirit of the novelist Kingsley Amis's anti-hero Lucky Jim, whose smoking habit was buoyed up by the thought that 'by the time I get cancer, they will have found a cure'. We can live in the hope of some future technological fix – maybe to escape from the planet altogether, and spread our destructive habits across the rest of the universe.

Yet what is dangerous is to euphemize any of these moves as 'sustainable', with its implicit message that the day of reckoning has been averted. The evasiveness of this phrase is deeply damaging to us. As George Orwell wrote in 1946, 'If thought corrupts language, language can also corrupt thought. A bad usage can spread by tradition and imitation, even among people who should and do know better.'

James Lovelock, who developed the Gaia hypothesis of the planet as a single self-regulating system, believes that, in the crisis of climate change, we should be thinking not of sustainable development but sustainable *retreat*. He has suggested that it might be necessary for humankind to draw back into the cities for a while to live in our own biomes on synthetically produced food, so that natural systems could have time to recover from the millennia of bludgeoning they've received at our hands. It's barely conceivable as a practical option, unless we experience a major catastrophe in the planet's life-support systems – a new Flood. And, at first sight, it seems in almost direct opposition to Eden's dream: a divorce, not a partnership, between people and plants. Yet maybe, as a concept, it's not so remote from Eden's project, simply putting human society inside a fenced enclosure instead of the plants, and making a gesture of real respect for the natural world.

At the moment we have no idea how to proceed. We are hopelessly ignorant about how ecosystems function to sustain life (including our own), and at a primitive stage in the development of an appropriate morality towards the rest of creation. Our attempts at a response – a small percentage drop in greenhouse emissions here, a little more financial support for environmentally friendly agriculture there – are well-meant, but little more than token gestures in the face of the enormity of change. We are frozen into inaction by a host of paradoxes and conflicting beliefs. We know that over-consumption is costing the earth, but are unwilling to collectively challenge it. We profess respect for the natural world and its ancient processes, but prefer our own solutions. We admire Eden's championing of 'useful' vegetation, but are reluctant to take that admiration to its logical extreme and accept that, perhaps, it is *all* useful, all an inextricable part of the whole.

Maybe we need to turn our conventional relationship with nature upside down, begin to learn *from* it rather than just 'about' it, let natural systems take the lead for once. Even the greenest of programmes are, normally, human projects modified to reduce their impact on the environment. What if we were to think the other way round instead, in terms of innovations that take natural forms and processes as their *models*, rather than their raw materials? Could we learn to 'grow' ceramics in the way that shellfish do? Create colours by the same gentle chemistry as flowers? Deal with dirt, not with synthetic detergents, however 'eco-friendly', but by developing paints and surfaces which are self-cleaning, as are so many leaves? Evolve a way of biologically generating energy that has the efficiency of photosynthesis? This is the agenda of the new science of biomimetics. It is in its infancy, and faces many obstacles. It is about listening to nature, not manipulating it, about a respect for the wild, not a commitment to its domestication, and thus seems to fly in the face of much conventional green wisdom. Or is it just another way of making an artefact of nature, the ultimate fence against the wild, inside which, as the most under-specialized organism on earth, we live with our prosthetic 'nature' and have no more need of the real thing?

The vegetable serpent?

IT IS IN SUCH CONFUSION THAT MYTHS GROW. They do not solve problems, but, as Lévi-Strauss suggests, help illuminate them, help 'resolve contradictions'. So here is my own fantasy for Eden, for an arena that might encourage the growth of a few. Down in those last fragments of the old clay pit, Eden should have a Wild Biome. Not another Wild Cornwall, artfully planted up to resemble the semi-wild communities of the area, but a space left utterly free of human interference, where wild plants and other organisms could work out their own development plans. It would be part of the Project by a single act of cultural interference: it would be labelled, signposted, so that it would be contemplated rather than passed over. Apart from that, it would remain literally and conceptually 'outside the fence'.

We can already guess what might happen to it from other clay pits, and from the tiny fragments of wild vegetation that have been allowed to cling on round the perimeter of the Project site. Seedling birch and rowan trees would rapidly invade, followed by oaks, and perhaps that ancestral forest tree of the south-west Cornish coast, the wild service-tree, whose fruits were one of England's aboriginal sources of sugar. And there would be furze, that squat, doughty, adaptable survivor that is the foundation of the English *garrigue*, and the signature plant of the whole Project.

Beyond that, what happened would be a matter of serendipity and surprise. Perhaps that rare and beautiful warbler of the south, the Dartford warbler, with its brilliant red eye and heather-pink breast, might return to nest in the furze. I have seen them some miles to the east, feasting on money-spiders dangled on gossamer threads out of the sky. But sooner or later, the furze would be joined by its companion plant, blown in on the wind. The mysterious dodder – leafless, rootless (heartless, some might say) – would twine its red lacework around the lower branches. Dodder is an epiphyte, like so many plants of the rainforest, but also a parasite. It has no attachments to the ground, and no chlorophyll. When its seeds germinate in late spring, they produce fine yellow threads, which later darken. The tip of the seedling is raised above the ground, and edges forward at the expense of the hindmost parts, spiralling all the while, so that it appears to be in slithering motion. In Cornwall, in an eerie echo of the vegetable

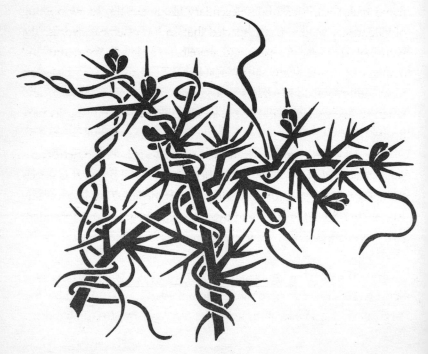

lamb, it was called Adder's Cotton. When the seedling meets a suitable host it begins to twine round it, alternating a series of loose coils with a series of tight ones. From these, tiny suckering spikes emerge, which penetrate the tissues of the host plant and provide a conduit for the absorption of water, food materials and enzymes. Sometimes when dodder attaches itself to small plants like thyme or an emergent heather, it will eventually kill its host. On furze, it is usually just a freeloader, weakening the plant a little where it has encircled it, but not fatally.

Dodder is of no direct practical use whatever to humankind. An earlier generation of herbalists used it as a purgative, because it appeared to have the signature of intestines in its sinuous stems, but of course it didn't work (though the belief is echoed in its specific Latin name, *epithymum*, Dioscorides' legendary laxative). They knew nothing of its parasitic nature, but suspected that, in John Gerard's words, 'the nature of this herbe changeth and altereth, according to the nature and qualities of the herbes whereupon it groweth'. And modern botanists have found that dodder does indeed exist in a number of races and types, each chemically adapted to recognizing and colonizing its particular host.

Dodder – clever, mobile, mutable, adaptable, living off other organisms without giving anything back because that is the way it is – feels strangely familiar. It has the makings of a myth about it, like the tiny weevil *symbiophilus*. If we could learn imaginatively to accept dodder's role, we might begin to acknowledge our own.

References, Sources and Notes

2   Gary Snyder, *Turtle Island*, 1974.

3   'the Genesis myth that reaches a climax': Rachel Storm, *Myths and Legends of the Ancient Near East*, 2003.

4   Theory of biogeography: Robert MacArthur, *The Theory of Island Biogeography*, 1967.

Lewis Thomas, *The Lives of a Cell*, 1974. ·

8   Claude Lévi-Strauss, *From Honey to Ashes*, 1973; Richard Mabey, *Flora Britannica*, 1996.

14   John Gerard, *The Herball, or Generall Historie of Plantes*, 1597, Dover Books facsimile of the 1633 edition, 1975.

16   For Borametz myths, see Henry Lee, *The Vegetable Lamb of Tartary*, 1887.

19   John Parkinson, *Paradisi in Sole Paradisus Terrestris*, 1629.

Lewis Thomas, op. cit.

21   For Middle Eastern myths, see Rachel Storm, op. cit.; Max Oelschlaeger, *The Idea of Wilderness*, 1991; Geoffrey Grigson, 'Ninhursaga' in *Gardenage*, 1952.

23–4   Desert food: Michael Abdalla, 'Wild growing plants in the cuisine of modern Assyrians in the eastern Syrian–Turkish borderland', Oxford Food Symposium, 2004; Don and Patricia Brothwell, *Food in Antiquity*, 1969.

26   John Berger, *Pig Earth*, 1979.

27   'written versions of Genesis': Robert Gould and Stephen Prickett (eds), *The Bible, Authorised King James Version*, 1997.

30   John Prest, *The Garden of Eden: The Botanic Garden and the Re-creation of Paradise*, 1981.

31   '"What had begun as a project"': Tim Smit, *Eden*, 2001; *The Lost Gardens of Heligan*, 1997.

39   For plant images in art, see N. K. Sandars, *Prehistoric Art in Europe*, 1985; Paul G. Bahn, *Journey through the Ice Age*, 1997; Wilfrid Blunt, *The Art of Botanical Illustration*, 1950.

41–3   Honey-hunting: David Lewis-Williams, *The Mind in the Cave*, 2002; Hattie Ellis, *Sweetness and Light: The Mysterious History of the Honey Bee*, 2004.

44   Walter Benjamin, *Illuminations*, English translation, 1970.

46   Lavenders: Tim Upson and Susyn Andrews, *The Genus Lavandula*, 2004.

51   For Mediterranean vegetation, see A. T. Grove and Oliver Rackham, *The Nature of the Mediterranean: An Ecological History*, 2001.

54   '"a sort of whip"': Oliver Rackham and Jennifer Moody, *The Making of the Cretan Landscape*, 1996.

55–8   History of lavender cultivation: Christiane Meunier, *The Lavender Country of Provence*, 1995; Tim Upson and Susyn Andrews, op. cit.

59   The Ruined Landscape hypothesis: A. T. Grove and Oliver Rackham, op. cit.

60   '"The Mediterranean forests..."': David Attenborough, *The First Eden*, 1987.

65   M. G. Morris and F. H. Perring (eds), *The British Oak*, 1974.

71   History of cereal farming: Don and Patricia Brothwell, op. cit.; Edward Hyams, *Plants in the Service of Man*, 1971.

72   History of bread: Jean Bottero, *The Oldest Cuisine in the World: Cooking in Mesopotamia*, 2002.

74–5   South American tribal myth: Claude Lévi-Strauss, op. cit.

76–7   Alternatives to cereals: J. Sholto Douglas and Robert A. de J. Hart, *Forest Farming*, 1976; N. Howes, *Nuts*, 1948.

78   Cultural history of olives: Mort Rosenblum, *Olives*, 1996.

83–4   Olives in Australia: Anne Dolamore, 'Where the Wild Things Are: From wild olives to present day cultivars', Oxford Food Symposium, 2004.

84–5   Olives in Tunisia: Mort Rosenblum, op. cit.

89   Edward O. Wilson, *Biophilia*, 1984.

97   'Smell is probably not': Diane Ackerman, *A Natural History of the Senses*, 1990.

99   Patrick Süskind, *Perfume*, 1986.

101   Avery N. Gilbert and Charles J. Wysocki, 'The Smell Survey', *National Geographic*, October 1987.

Scent as agent of memory: Stephen Harrod Buhner, *The Lost Language of Plants*, 2002; Brian J. Ford, *Sensitive Souls: Senses and Communication in Plants, Animals and Microbes*, 1999.

105   Doctrine of Signatures: E. S. Rohde, *The Old English Herbals*, 1972; Thomas Hill, *The Gardener's Labyrinth*, 1593, Oxford edition edited Richard Mabey, 1988.

106   Anthony Cavender, *Folk Medicine in Southern Appalachia*, 2003.

107   Ginseng: Andrew Dalby, 'Ginseng: Taming the Wild', Oxford Food Symposium, 2004; Euell Gibbons, *Stalking the Healthful Herbs*, 1966.

110   *Matti ka attar*: Magnus Pyke, 'Smells Like Rain', BBC Radio 3 talk, October 1971.

116   John Prest, op. cit.

117   S. E. Morison, *Admiral of the Ocean Sea: A Life of Christopher Columbus*, 1942.

118 Sugar, and ecological history of Madeira: Alfred W. Crosby, *Ecological Imperialism: The Biological Expansion of Europe, 900–1900*, 1986.

120 Sugar plantations: Toby and Will Musgrave, *An Empire of Plants*, 2000; Henry Hobhouse, *Seeds of Change*, 1999.

123 Honey: Hattie Ellis, op. cit.; Bee Wilson, *The Hive: The Story of the Honeybee and Us*, 2004.

126–7 Claude Lévi-Strauss, op. cit.

127–8 Honey myth: This is my own rendering of the myth.

130 Sugar beet: Joan Thirsk, *Alternative Agriculture: A History*, 1997.

135 Use of hemp: Anna Lewington, *Plants for People*, 2003.

137–8 For hemp in East Anglia, see Arthur Young, *A General View of the Agriculture of Norfolk*, 1804; Eric Pursehouse, *Waveney Valley Studies*, 1966; Michael Friend Serpell, *A History of the Lophams*, 1980.

141 For translocation of plants, see Lucile H. Brockway, *Science and Colonial Expansion: The Role of the British Royal Botanic Gardens*, 1979; Ray Desmond, *Kew: The History of the Royal Botanic Gardens*, 1995.

Wardian Cases: Nicolette Scourse, *The Victorians and their Flowers*, 1983; Lynn Barber, *The Heyday of Natural History*, 1980.

142–3 Shirley Hibberd, *Rustic Adornment for Homes of Taste*, 1870.

143–4 Orchid hunting: Tyler Whittle, *The Plant Hunters*, 1970.

147 Joseph Dalton Hooker, *Himalayan Journals*, 1855.

148 Eric Hansen, *Orchid Fever*, 2000.

156 Eleazar Albin, *A Natural History of British Song-birds*, 1737.

157–8 Edward O. Wilson, *The Diversity of Life*, 1992.

163 The Amazonian water-lily: Wilfrid Blunt, *In for a Penny: A Prospect of Kew Gardens*, 1978.

167 Ray Desmond, op. cit.

168 For the Titan, see account in *Curtis's Botanical Magazine*, Vol. 117, 1891.

170–1   Hogweed: William Robinson, *The Wild Garden*, 1870.

172   Richard Mabey, *The Flowering of Kew*, 1988.

173   Mary-Louise Pratt, *Imperial Eyes: Travel Writing and Transculturation*, 1992.

174–7   Francis Masson's journal in *Philosophical Transactions*, 1776.

177–81   Company art: see Richard Mabey, op. cit., 1988; Mildred Archer, 'India and Natural History: the role of the East India Company', *History Today*, November 1959; P. Hulton and L. Smith, *Flowers in Art from East and West*, 1979.

181–5   Maria Merian: Natalie Zemon Davis, *Women in the Margins*, 1995; Madeleine Pinault, *The Painter as Naturalist*, 1991.

185–8   Margaret Mee, *In Search of Flowers of the Amazonian Forest*, 1988.

186   Geoffrey Grigson (ed.), *Thornton's Temple of Flora*, 1972.

190   Swamp Circus are at www.swampcircus.com

191–3   Rainforest canopy: Adrian Bell, 'On the roof of the rainforest', *New Scientist*, 2 February 1991; Stephanie Pain, 'Over the Forest of Bees', *Kew Magazine*, Spring 2000.

194   Krondenak are at www.treemail.nl/krondenak/cic.htm

200   Ideas about the restoration of the Garden of Eden: Jim Bennett and Scott Mandelbrote, *The Garden, the Ark, the Tower, the Temple: Biblical Metaphors of Knowledge in Early Modern Europe*, 1988.

202   John Evelyn, *Acetaria: A Discourse of Sallets*, 1699.

204   Tim Smit, *Eden*, 2001.

205   '"An architect would fall"': Martin Jackson, *Eden: The first book*, 2001.

212   '"social justice"': Jared Diamond, *Collapse: How Societies Choose to Fail or Survive*, 2005.

216   George Orwell, 'Politics and the English Language', 1946, in

Collected Essays, Journals and Letters, Vol. IV, 1968.

218   For biomimetics: Janine Benyus, 'Genius of Nature', *Resurgence,* May/June 2005. Also www.bioneers.org

Acknowledgements

My warmest thanks for their help and encouragement to the whole Eden Team, from gardeners to gurus. Since, as in the best of working communities, they repeatedly change from one to the other, it would be invidious to pick anyone out; to this outsider they seem to have melded with the Eden landscape every bit as successfully as the biome robins. The views expressed in the book, though, are my own, not the Project's, though I sense we have a lot of common ground.